Best Regards, Annette
Feb. 2008

Demographic Research Monographs

A Series of the Max Planck Institute
for Demographic Research, Rostock, Germany

Editor-in-chief: James W. Vaupel

Demographic Research Monographs

E. Barbi, S. Bertino, E. Sonnino (Eds.)
Inverse Projection Techniques
2004. XI, 137 pages. 978-3-540-20931-7

G. Doblhammer
The Late Life Legacy of Very Early Life
2004. X, 204 pages. 978-3-540-22105-0

R. Rau
Seasonality in Human Mortality
2007. XV, 214 pages. 978-3-540-44900-3

Annette Baudisch

Inevitable Aging?

Contributions to Evolutionary-Demographic Theory

Springer

Dr. Annette Baudisch
Max Planck Institute for Demographic Research
Konrad-Zuse-Straße 1
18057 Rostock
Germany
baudisch@demogr.mpg.de

Printed with the financial support of the Max Planck Society

ISBN 978-3540-76655-1 e-ISBN 978-3-540-76656-8

DOI 10.1007/978-3-540-76656-8

Demographic Research Monographs ISSN 1613-5520

Library of Congress Control Number: 2007940154

Production: LE-TEX Jelonek, Schmidt & Vöckler GbR, Leipzig
Cover-design: WMX Design GmbH, Heidelberg
The front cover photo – top right hand corner – includes a photograph by Harald Wenzel-Orf (www.wenzel-orf.de) and has been reproduced with his permission.

Printed on acid-free paper

9 8 7 6 5 4 3 2 1

springer.com

To my family

Foreword

What is the ultimate cause of aging? Before Dr. Annette Baudisch began her research on this question, aging was thought to be fundamentally a consequence of the decline with age of the force of Darwinian selection. Only a fraction of individuals survive to older ages and only a fraction of their fertility remains. Consequently, deleterious mutations that act only at older ages will be purged so slowly over successive generations that such mutations will accumulate in the population. This implies that starting at the age of reproductive maturity death rates will begin to increase. As William D. Hamilton put it, "senescence is an inevitable outcome of evolution." This view was biological dogma for half a century – until Dr. Baudisch's compelling critique and radical breakthrough, cogently explained in this monograph.

Dr. Baudisch distinguishes between aging and senescence, defining aging as the pattern of change in mortality over age and senescence as an increase in mortality with age. She shows that for some species mortality goes down with age or stays constant – something she calls sustenance. In this monograph she presents some empirical evidence for this, but mainly she develops the theoretical argument that the age-pattern of mortality is a consequence of a species' optimizing limited resources. Tradeoffs must be made between spending energy on growth, repair and maintenance, on the one hand, and reproduction, on the other. If enough is spent on growth, repair and maintenance, then there is no increase in mortality with age. Senescence results from the cumulative impact of an imbalance between damage and repair. If, as Dr. Baudisch demonstrates, the level of repair and regeneration is sufficient to counter-balance new damage and destruction, then an organism can maintain itself. With this research Dr. Baudisch has made a major contribution to knowledge and has created a new agenda for research on the evolutionary biology of aging.

The first chapter of Dr. Baudisch's monograph sets the stage by reviewing evolutionary theories of senescence, discussing how senescence and fitness can be measured, outlining the optimal life-history approach, and summarizing some recent research developments. The chapter is much more than a critical survey of the literature: it is a lucid, original account of key research questions and strategies.

The second chapter carefully considers Hamilton's indicators of the force of selection. Insight after insight is gained as Dr. Baudisch probes deeper and deeper. She shows that Hamilton's claim that senescence is inevitable can be disproved even within Hamilton's restricted framework. In particular, she demonstrates that depending on the measure used and on levels of mortality and fertility, the force of selection can increase with age. Furthermore, she explains why and how the age-trajectories of mortality and fertility depend not only on the force of selection but also on the incidence of mutations. Dr. Baudisch's findings in this chapter strengthen the case that demographic schedules of mortality and fertility over most of the life course are largely shaped by optimization of tradeoffs rather than by mutation accumulation. An earlier version of the chapter was published as a single-authored article[1] in *PNAS, the Proceedings of the National Academy of Sciences* of the United States.

In her third chapter, Dr. Baudisch critically explains the limitations of Hamilton's framework. By presenting empirical evidence and by systematically reviewing various theoretical considerations, she develops further reasons why mutation accumulation is of secondary importance in molding age-trajectories of mortality. Instead, the primary force is adaptation: patterns of aging are a byproduct of the optimization of tradeoffs. In addition, Dr. Baudisch shows that Hamilton's notion – that the age pattern of mortality is inversely related to his indicator – is wrong. The relationship between the force of selection and the pattern of mortality is so complicated that sophisticated modeling is required.

Chapters 4 and 5 of Dr. Baudisch's monograph develop two kinds of optimization models, which she analyzes using sophisticated methods of control theory and dynamic programming. In Chapter 4 the focus is on the role of growth. By investing in continued growth, the individuals

[1] Annette Baudisch. Hamilton's Indicators of the Force of Selection. *Proceedings of the National Academy of Sciences, USA*, 102(23):8263–8268, 2005. Copyright (2005) National Academy of Sciences, U.S.A.

in a species can attain greater size. If chances of survival increase as size increases, then death rates can fall with continued growth. Dr. Baudisch shows that it can be optimal to grow rather than to invest more available resources in fertility.

In Chapter 5, she considers a model of vitality, i.e., size adjusted for functioning. Dr. Baudisch shows that various patterns of aging are optimal depending on a species' characteristics. In particular, sometimes senescence is optimal and sometimes it is not. The dichotomy depends on various factors, including the hazardousness of the environment, whether there are economies of scale in investing in growth and maintenance, and whether there are economies of scale in reproduction.

Only under particular conditions do Dr. Baudisch's models in Chapters 4 and 5 lead to senescence. Her simpler models imply sustenance; senescence is optimal only in more complicated, restricted models. Not only is senescence not inevitable: senescence is a special case. In her sixth and final chapter, Dr. Baudisch points to directions for future research. Some of the suggestions concern the burning questions she would like to address next. Other suggestions pertain to the development of the field of evolutionary demography. This is an exciting, insightful chapter, full of ideas presented in a judicious balanced manner. It will guide other researchers: it adumbrates an important new field of inquiry.

The monograph builds on Dr. Baudisch's doctoral dissertation. In particular, Chapter 5 has been substantially revised and strengthened, in part of a result of an insight of Arthur Robson, Professor of Economics at Simon Fraser University, and comments from and subsequent interchanges with Kenneth Wachter, Professor of Demography and Mathematical Statistics at the University of California Berkeley, member of the U.S. National Academy of Sciences. Prof. Linda Partridge, an evolutionary biologist at University College London who was recently appointed Director of the new Max Planck Institute for the Biology of Aging, and I served as her dissertation advisors. Annette Baudisch received her doctorate after being examined in June 2006 by Michael Murphy, Professor at the London School of Economics and past President of the British Population Association and Brian Charlesworth, Professor at the University of Edinburgh and Fellow of the Royal Society, an evolutionary biologist who has focused much of his research on evolutionary theories of aging.

Kenneth Wachter emailed Dr. Baudisch: "I read your dissertation... It is the most thought-provoking and comprehensive dissertation I have read in many years." This judgment was confirmed when the Max Planck Society in June of this year awarded Dr. Baudisch the Otto Hahn Medal, the Society's highest prize for doctoral research.

The pathbreaking research of Dr. Annette Baudisch in her dissertation, further developed in this monograph, fundamentally alters understanding of the evolution of aging. Numerous researchers will spend years exploring the field she has opened up, trying to understand why some kinds of species suffer senescence whereas others enjoy sustenance. Her dissertation is exceptionally important, stimulating and insightful. This monograph is even better.

The series of Demographic Research Monographs is under the editorial supervision of the Max Planck Institute for Demographic Research. I am Editor-in-Chief. I am advised by an Editorial Board that currently consists of Prof. Elisabetta Barbi (Messina University, Italy), Prof. Gabriele Doblhammer (Rostock University, Germany), Dr. Jutta Gampe (Max Planck Institute), Prof. Joshua Goldstein (Max Planck Institute), and Prof. Bernard Jeune (University of Southern Denmark). Additional members of the Editorial Board will be appointed as needed to review manuscripts submitted for possible publication. The current manuscript was reviewed and accepted by James Vaupel, Joshua Goldstein and Jutta Gampe.

The Demographic Research Monographs series can be considered the successor to the series called Odense Monographs on Population Aging, edited by Bernard Jeune and James Vaupel. The volumes in this now-terminated series were first published as hardcover books by an academic publisher, Odense University Press, and subsequently made available online at www.demogr.mpg.de/books/odense. The nine Odense Monographs on Population Aging include two collections of research articles that focus on specific subjects on the frontier of demographic research, three volumes by senior researchers that present path-breaking findings, a review of research on a topic of emerging interest, a presentation of a new method for analysis of demographic data, an outstanding doctoral dissertation, and a unique collection of important demographic data on non-human species.

The series of Demographic Research Monographs will continue this mix, with books that are often under 200 pages in length, that have a clear focus, and that significantly advance demographic knowledge. Research related to population aging will continue to be a focus on the series, but it will not be the only one. The series will embrace all of demography, broadly defined. As indicated by the first volume in the series, an important subject will be historical demography. We also plan to publish research on fertility and family dynamics. Mathematical demography is the core of the population sciences and we will strive to foster monographs that use mathematics and statistics to further develop the theories and methods of demography. Biodemography is a small but rapidly growing and particularly innovative branch of demography: we will seize opportunities to publish monographs at the intersection of biology and demography, pertaining both to human and other species, and including demographic research with ties to such fields as epidemiology, genetics, evolutionary biology, life-history biology, experimental demography, and paleodemography.

Each volume in the Demographic Research Monograph series will have a substantial link to the Max Planck Institute for Demographic Research. As well as being published as hardcover books by Springer-Verlag, the volumes of the Max Planck series of Demographic Research Monographs will subsequently be available at www. demogr. mpg. de /books/drm. The online version may include color graphs, supplemental analyses, databases and other ancillary or enhanced material. Parallel publication online and in print is a significant innovation that will make the monograph series particularly useful to scholars and students around the world.

Rostock,
August 2007

James W. Vaupel
Editor-in-Chief

Preface

The research presented in this monograph would not have been possible without the substantial support from many people. I am especially grateful to my advisors Linda Partridge at University College London and James W. Vaupel at the Max Planck Institute for Demographic Research. I was fortunate that they gave me the freedom to explore my intellectual interests while providing continuous support and advice. Linda and, even more, Jim often challenged me through detailed and intensive discussions.

I am deeply grateful to the Max Planck Institute for Demographic Research for a variety of reasons. The administrative and technical staff at the institute provided superb research support. Frequent traveling to workshops and conferences encouraged and supported by the Max Planck Institute gave me the chance to exchange ideas with many established researchers in my field of interest. The International Max Planck Research School of Demography provided comprehensive doctoral training. My colleagues created a stimulating, enjoyable and friendly working atmosphere at the institute. I especially wish to thank Jutta Gampe for all her support and Roland Rau for patiently helping me with whatever question I had.

I further benefited from being part of the Research Training Network created by Europe's leading demographic research centers with financial support from the European Commission. The network enabled me to spend 4 months at the Vienna University of Technology carrying out my research and receiving training in dynamic optimization and demography under the supervision of Gustav Feichtinger and Alexia Fuernkranz-Prskawetz.

Wherever I went, I was lucky to meet friendly, open-minded and collaborative people who helped me to find my way. I thankfully acknowledge guidance from Kenneth Wachter and Brian Charlesworth. I am grateful to Brian Charlesworth and to Michael Murphy for serving as my doctoral examiners. I have also benefitted from helpful discussions with Peter Abrams, James Carey, Hal Caswell, Kaare Christensen, James Curtsinger, Patrick Doncaster, David Gems, Ronald Lee, Daniel Martinez, Anatoli Michalski, Steven Orzack, Daniel Promislow, Deborah Roach, Arthur Robson, David Steinsaltz, Francois Taddei, Marc Tatar, Shripad Tuljapurkar, Vladimir Veliov and many others.

Finally, I owe a great debt to my family and my friends for their love and support.

Rostock,
August 2007

Annette Baudisch

Contents

Part II Optimization Models

List of Figures

List of Tables

1

Introduction

1.1 Synopsis

Death is part of life, and it can strike any time. The question is whether death necessarily becomes more likely as life proceeds. William D. Hamilton (1966), one of the leading biologists of the last century claimed that senescence is inevitable[1] because the force of selection declines with age, making later ages unimportant to evolution. Survival and reproduction are the key players in this game and they are the traits negatively affected when selection loosens its grip.

Since 1966 it has been dogma among gerontologists that a decline in physiological functioning with age, i.e. senescence, is an inherent, inescapable part of life. Humans inevitably grow old, which is probably why it seems so unlikely to us that other forms of life could escape senescence. Biologists, however, often observe that functioning improves as individuals develop. Therefore the idea of living beings that perform equally well or better over their life course until they eventually meet the Grim Reaper might not be so strange after all.

One major result of my article published in *PNAS* [11] is that no dogmatic statement can be made about the universality of senescence. By carefully studying Hamilton's work on the molding of senescence I show that Hamilton did not prove that senescence is unavoidable. He claimed that the force of selection must decrease with age for any

[1] The word "aging" is often used instead of the narrower, more precise but less common word "senescence" to describe a decline in physiological functioning with age. Hence I chose to entitle this monograph "Inevitable aging?" instead of "Inevitable senescence?". Throughout the monograph, however, I make a clear distinction between aging and senescence: I use the term aging to refer to any kind of variation in functioning with age, for the better or worse, and reserve the term senescence for a deterioration in functioning.

conceivable organism. The weaker the force of selection, the more unfavorable mutations might sneak in, constituting a mutational burden. Contrary to his results, I point out that the force of selection can increase with age and, in this case, will counteract mutational burden at higher ages more strongly than at younger ages. The specific nature of a mutational effect, i.e. whether a mutation affects mortality in an additive or in a proportional way, determines the dynamics of the force of selection with age.

Combining Hamilton's analysis with the concept of mutation– selection balance and providing a critical analysis of theoretical issues and empirical evidence, I strengthen the view that the age-patterns of mortality and fertility are largely shaped by optimization rather than by the accumulation of deleterious mutations. However, the question of the impact of mutational burden vs. optimization is not yet closed.

Building on the insight that senescence is likely to be a byproduct of an adaptive process, I developed simple state-dependent models, three based on size and one on vitality.

The size-based models [200] show that negative senescence can be an optimal life-history strategy. The trajectory of growth is a crucial determinant in tipping the scale between senescence and sustenance. Indeterminate growers, i.e. species that exhibit a period of parallel growth and reproduction as part of their life history, are likely candidates for sustenant strategies, whereas senescence is expected for species that stop growing at about the age of reproductive maturity.

A fundamental insight gained from the vitality-based optimization approach, vitality being the size of an individual weighted by functioning, is the major importance of the costs of maintenance and growth for the determination of senescence versus sustenance. The model shows that a rich diversity of age-patterns of mortality can be optimal. Sustenance outperform senescence when maintenance costs are low. I show that changes in intrinsic and extrinsic mortality can switch the life history between senescence and sustenance strategies if the level of costs of reproduction and growth is not too high. The model is a step forward in identifying the characteristics in a species that predict whether the species follows a senescent or a non-senescent life history.

A further insight from the vitality model concerns a mortality paradox. Contrary to "Williams' Hypothesis" that species living under more hazardous extrinsic conditions should exhibit faster senescence, I show that an increasing extrinsic hazard could switch an optimal life history from a senescent to a non-senescent one if maintenance costs are low.

In all my models, optimal equilibrium is assumed, something that might never be reached in nature. The variability of the environment is neglected. Competition between individuals in a population and among populations as well as the resulting interdependent population dynamics are not taken into account. One might perhaps claim that I study evolution without evolution. I defend my approach with the argument that I wish to study whether and when senescence can be avoided by any conceivable organism. The idea is that if senescence is not inevitable and is only one of many options for the age-patterns of life in optimal equilibrium, then this is a hint that the real world may provide these options as well.

1.2 Background

1.2.1 Senescence – Paradox? – Inevitable?

Life is shaped by evolution as described by Darwin [48, p. 5]:

> "As many more individuals of each species are born than can possibly survive; and as, consequently, there is a frequently recurring struggle for existence, it follows that any being, if it vary however slightly in any manner profitable to itself, under the complex and sometimes varying conditions of life, will have a better chance of surviving, and thus be *naturally selected.* From the strong principle of inheritance, any selected variety will tend to propagate its new and modified form".

The key players in evolution are survival and reproduction. To reproduce you have to be alive, to be selected you need to reproduce more successfully than your competitors, and finally you have to transmit this ability to your offspring. Senescence is a process of decline in physiological functioning that results in a decrease in survival and/or reproduction with age. Therefore, senescence is an unfavorable process in the struggle for existence. The question arises: Why, then, could it evolve at all? Clearly, senescence did evolve – but did it evolve in all forms of life? This is the burning question I wish to answer from a theoretical perspective. Is senescence an inherent part of life or could it be that some species have escaped senescence?

William D. Hamilton wrote a very influential article in 1966 on "The moulding of senescence by natural selection," in which he claimed that senescence is inevitable. Hamilton states that "no life schedule, even under the most benign ecology imaginable, could escape my spectrum of forces of senescence . . . in the farthest reaches of almost any bizarre universe" [76, p. 90]. "[F]or organisms that reproduce repeatedly, senescence is to be expected as an inevitable consequence of the working of natural selection" [76, p. 109]. Did Hamilton really prove that senescence is inevitable? I will treat this question in Chaps. 2 and 3, and the answer is: No, he did not.

1.2.2 Evolutionary Theories of Senescence

Two main approaches have been developed to explain the evolution of senescence: The first approach assumes that senescence is due to a burden of deleterious mutations at later ages, whereas the second approach assumes that senescence is a negative byproduct of an adaptive process constrained by trade-offs. Both approaches hinge on the assumption that the force of selection declines with age. The force of selection is determined by differences in reproductive success. The larger the difference in reproductive success between two alternative variants of a trait, the stronger the force of selection on that trait. Reproductive success is determined by survival and reproduction. Consequently, the force of selection is determined by survival and reproduction.

Since death is certain, the number of survivors of a birth cohort declines with age. Medawar [126] conjectured that, because fewer and fewer individuals survive up to higher and higher ages, those ages matter less and less to life-time reproductive success, leading to a decline in the force of selection with age. Hamilton [75] thought he had proved that the force of selection must decline with age, but I will show later that, under some circumstances, the force of selection can increase with age.

Medawar [126] proposed the theory of mutation accumulation. Mutations occur recurrently. To the extent that reproduction or survival are in any way negatively affected, an individual carrying such a mutation will be at an evolutionary disadvantage relative to non-carriers of that mutation. Clearly, the force of selection would tend to wipe out deleterious mutations. However, as the force of selection peters out, bad mutations manage to creep in, being less and less strongly opposed by evolutionary forces. Medawar argues that the smaller the force of selection, the more mutations would accumulate.

Williams [212] proposed the theory of antagonistic pleiotropy after the basic idea was initially formulated by Medawar [126, p. 64]. Like the theory of mutation accumulation, Williams's approach is based on the precondition that the force of selection decreases with age. Genes are considered that have fitness enhancing effects earlier in life and fitness depressing effects later in life. Because the force of selection decreases with age, the advantage early in life receives a much stronger weighting than the disadvantage late in life. Unlike the passive process underlying mutation accumulation, mutations are actively selected that imply a deleterious effect at older ages, since the balance between costs and benefits favors younger ages.

Note that the general idea underlying antagonistic pleiotropy is to actively balance linked traits that affect survival and reproduction in opposite ways. Genes with antagonistic and pleiotropic effects are a specific case of a trade-off affecting fitness. The general idea of trade-offs underlies the disposable soma theory proposed by Thomas Kirkwood [97, 98]. Kirkwood's approach is based on the observation that the critical part of an individual that must survive is the genetic code. The genetic code contains all information needed to ensure the persistence of a lineage. It is therefore economic to separate the germ cells from the rest of the body cells, the soma, and to protect only the germ line from the ubiquitous occurrence of damage. The soma merely serves as a vehicle for the genetic code to be transported over generations. Kirkwood conjectured that the costs required for the persistent repair of the soma is too high and evolution therefore trades off the protection of the germ line against senescence of the soma.

1.2.3 Measuring Senescence

Senescence can be defined as a decline in physiological functioning with age that negatively affects the ability to survive and/or to reproduce. There is, however, no generally agreed upon measure of senescence.

One approach to measure senescence is to look at the change in mortality with age. In this case, senescence corresponds to an increase in mortality with age. This is a simple and widely accepted working definition [56, p. 12].

Since mortality and fertility are closely linked, an ultimate measure of senescence should include both survival and reproduction. Partridge and Barton [149] suggest using reproductive value at age a to determine the state of senescence of an individual. Reproductive value captures the remaining reproductive contribution of an individual that is alive at age a. It was defined by Fisher [59] as

$$v(a) = \frac{e^{r\,a}}{l(a)} \int_a^\infty e^{-r\,x}\, l(x)\, m(x)\, dx \,. \tag{1.1}$$

The survival function $l(x)$ indicates the probability of survival from birth (or conception) to age x and the maternity function $m(x)$ indicates age-specific reproduction. Age-specific survival and reproduction are weighted by the population growth term $e^{-r\,x}$, which discounts future reproduction by the intrinsic rate of population increase r [94]. The integral sums up all reproductive contributions from age a onwards. Multiplication by $e^{r\,a}/l(a)$ accounts for the fact that the individual has already survived to age a.

Senescence in this framework corresponds to cases when reproductive value declines with age, i.e. the derivative of $v(a)$ given in (1.1) with respect to age is negative,

$$\frac{dv(a)}{da} < 0\,. \tag{1.2}$$

Applying the product and chain rules from basic calculus yields

$$\begin{aligned}\frac{dv(a)}{da} = r\,\frac{e^{r\,a}}{l(a)} \int_a^\infty \; & e^{-r\,x}\, l(x)\, m(x)\; dx \\ & - \frac{e^{r\,a}}{l^2(a)}\,\frac{dl(a)}{da} \int_a^\infty \; e^{-r\,x}\, l(x)\, m(x)\; dx \\ & \qquad - \frac{e^{r\,a}}{l(a)}\, e^{-r\,a}\, l(a)\, m(a) \, < \, 0\,. \end{aligned} \tag{1.3}$$

Note that the probability of survival to age a, $l(a)$, is determined by the age-trajectory of mortality $\mu(x)$ from age zero to age a through the relation

$$l(a) \;=\; e^{-\int_0^a \,\mu(x)\; dx}\,. \tag{1.4}$$

Thus, (1.3) can be simplified by substituting

$$\mu(a) \;=\; -\frac{\frac{dl(a)}{da}}{l(a)} \tag{1.5}$$

as well as substituting expression (1.1) for reproductive value, which leads to

$$\frac{dv(a)}{da} \;=\; r\,v(a) \,+\, \mu(a)\,v(a) \,-\, m(a) < 0\,. \tag{1.6}$$

After rearranging it can be concluded that senescence occurs when

$$v(a) \;<\; \frac{m(a)}{\mu(a) + r}\,, \tag{1.7}$$

where $\mu(a) + r > 0$. Note that, if mortality and fertility do not change with age , i.e. $m(a) = m$ and $\mu(a) = \mu$, then – following from its definition in 1.1 – reproductive value is constant at the level

$$v(a) = \frac{m}{\mu + r} \tag{1.8}$$

for all ages a. Conditions (1.7) and (1.8) imply that senescence occurs if reproductive value at age a is lower than it would be if both mortality and fertility remained constant from that age onwards. Clearly if mortality and fertility are constant, then the organism does not senesce. Condition (1.7) implies that at least one of the two fitness components is adversely affected, which is intuitively appealing.

The change in reproductive value with age accounts for both the change in mortality and fertility, which is a favorable argument for its use as a measure of senescence. However, reproductive value in general and condition (1.7) in particular take into account the whole remaining life history. It seems more reasonable that the state of senescence of an individual at a certain age interval should be determined by changes in mortality and fertility at that specific age interval alone without any knowledge about the future. Furthermore, note that the population growth rate r enters the measure of senescence if reproductive value is used to account for the senescent state of an individual. But why should the population growth rate influence the definition of senescence? This issue disappears under the optimal equilibrium assumption since $r = 0$.

An alternative definition of senescence can be derived that accounts only for changes in the state of an individual at the current age interval, determined by mortality and fertility. Senescence corresponds to cases where mortality increases while reproduction is constant or decreases with age. Senescence also occurs if mortality does not change with age but fertility decreases. On the other hand, no senescence is observed if mortality decreases or remains constant and fertility increases or remains constant.

If mortality and fertility both increase, or both decrease, one has to be careful. If, for instance, fertility increases but mortality increases even more, then the loss in survival outweighs the gain in reproduction. If, on the other hand, mortality decreases, say, at a rate of -2% but fertility decreases even more, say, at a rate of -4%, then the gain in survival is more than erased by the loss in reproduction, i.e. $-4\% < -2\%$. In sum, senescence depends on the change in mortality vs. the change in fertility.

Formally, this can be expressed by comparing the relative change in mortality with the relative change in fertility. Relative changes are used to produce comparable quantities with the same units; change per time. The relative change in mortality is given by

$$\frac{\frac{d\,\mu(a)}{da}}{\mu(a)} \equiv \acute{\mu}(a) \; , \tag{1.9}$$

where the change in mortality over age relative to the current level of mortality is denoted by the short hand notation $\acute{\mu}(a)$. The same holds analogously for fertility $m(a)$.

In general, senescence[2] pertains to cases when the relative change in mortality is greater than the relative change in fertility at age a, i.e.

$$\acute{\mu}(a) > \acute{m}(a) \; . \tag{1.10}$$

Table 1.1 summarizes the cases for senescence vs. non-senescence [3].

Table 1.1. Senescence or not

	$\acute{m}(a) > 0$	$\acute{m}(a) = 0$	$\acute{m}(a) < 0$
$\acute{\mu}(a) > 0$	sen if $\acute{\mu}(a) > \acute{m}(a)$ not if $\acute{\mu}(a) \leq \acute{m}(a)$	sen	sen
$\acute{\mu}(a) = 0$	not	not	sen
$\acute{\mu}(a) < 0$	not	not	sen if $\acute{\mu}(a) > \acute{m}(a)$ not if $\acute{\mu}(a) \leq \acute{m}(a)$

The burning question of my work is whether the lower "triangle" in Table 1.1 is filled with life. Are there life histories that lack senescence which have been evolutionarily more successful than life histories with senescence? The first step on the way to answering this question is to

[2] Note that my definition of senescence is a demographic definition, i.e. on the level of changes in mortality and fertility. The definition of senescence as decline in physiological functioning (see [172]) pertains to the level of phenotypic traits. It is possible that some changes in physiology do not become apparent (at least not immediately) at the demographic level.

[3] Carey and colleagues [21] point out that mortality patterns of medflies fluctuate up and down with age, which would correspond to "alternating periods of positive and negative senescence. It is questionable whether it is helpful to define the word senescence in this way." I agree that short-term fluctuations in mortality may not indicate positive vs. negative senescence. Consequently, in defining senescence as in (1.10), it is important to consider changes in mortality and fertility over reasonable age intervals, which should be determined relative to a species' lifespan.

determine how to measure "fitness", i.e. the evolutionary success of a strategy.

1.2.4 Measuring Fitness

The notion "fitness" captures the reproductive success of a genotype. Reproductive success results in population growth. Fitness is therefore often measured by the intrinsic rate of population increase, r, which is implicitly defined by the Lotka Equation [179],

$$1 = \int_0^\infty e^{-r\,a}\, l(a)\, m(a)\, da \,. \tag{1.11}$$

From the beginning of life until the end, this integral sums up age-specific reproduction $m(a)$, which can only be realized if an individual is alive at age a, captured by $l(a)$. Furthermore, later-born offspring are discounted by population growth ($e^{-r\,a}$) because earlier-born offspring contribute relatively more to future generations. The value of r that uniquely satisfies this equation for given schedules of $l(a)$ and $m(a)$ is the intrinsic rate of population increase.

Another frequently used measure of fitness is the net reproduction rate, R, given by

$$R = \int_0^\infty l(a)\, m(a)\, da \,. \tag{1.12}$$

Note that R counts the number of offspring produced per lifetime, accounting for survival. This measure of fitness is appropriate when the population size does not change. Otherwise, the intrinsic rate of population increase is more appropriate.

Both fitness measures hinge on the underlying assumptions of stable population theory. In his famous equation Lotka assumes a homogeneous population that is closed to migration. Either individuals are of one sex or individuals of only one sex determine r and R. Birth and death rates are constant over time and the environment is unchanging. There are no density effects. Intergenerational transfers such as parental care are neglected.

In the 1970s Charlesworth, building on Haldane [72] and Norton [141], justified the use of r as a fitness measure. The results of Charlesworth [24] show that in an age-structured, diploid, randomly mating population r can be associated with the fate of a rare, nonrecessive gene. In Charlesworth [25] he gives approximations that are otherwise necessary. A comprehensive treatment can be found in Charlesworth [27, Sect. 4.6.1].

The use of the intrinsic rate of population increase, r, is accepted as a reasonable working assumption [27, 28, 172] for cases of constant and density-independent environments, but one must be aware of its restrictions (see Chap. 6).

1.2.5 Optimal Life History

An optimal life history is captured by the age-trajectories of survival and reproduction that maximize fitness. Fitness can be measured by the intrinsic rate of population increase r and is determined by the schedules of survival and reproduction. In this context it is important to highlight that optimal life-history schedules depend on the level of r [69]. If a population grows quickly, later births are devalued heavily and therefore a short generation time are favored. This strategy might differ substantially from a strategy that maximizes fitness in a non-growing, stationary population.

In my work, I will assume a population that is in long-term optimal equilibrium. I will not consider the evolutionary process of getting there and I will exclude the possibility that an equilibrium might never be reached. This is a simplified but reasonable assumption because, on an evolutionary time scale, any small deviation from $r = 0$ will have strong consequences: *"...any being, if it vary however slightly in any manner profitable to itself, under the complex and sometimes varying conditions of life, will have a better chance of surviving, and thus be naturally selected."* [48, p. 5]. Many species have survived in essentially unchanged form for many generations: their life histories may be close to optimal. In any case, it is possible that some species are close to optimal equilibrium and it is of interest to study whether for such species senescence is inevitable. If it is, then this strengthens Hamilton's case. If it is not, this disproves Hamilton's claim that senescence is inevitable for any conceivable organism.

Taylor and colleagues [192] analytically proved that "*[m]aximizing the reproductive value at age zero is mathematically equivalent to maximizing the ultimate rate of increase*". Here r is referred to as the ultimate rate of increase in order to emphasize that this is the rate to which a population's growth rate will ultimately converge [94]. Discussion of the theorem was raised by Caswell [22], who claimed that this would hold only under some very specific conditions. Yodzis [219] clarified the issue and showed that Taylor and colleagues [192] were generally right. However, he also pointed out the critical restrictions. First, maximizing the reproductive value gives only a local maximum of r. Second, the use of r as a fitness measure is an issue in itself. And third, the consequences of population regulation mechanisms, such as predation and density effects, are not taken into account.

For $r = 0$ the reproductive value given in (1.1) at age $a = 0$ equals the net reproduction rate R given in (1.12), which is an alternative measure of fitness to r (see Sect. 1.2.4). Following the result of Taylor et al. [192] maximizing R is equivalent to maximizing r such that $r_{max} = 0$.

Maximizing life-time reproduction R with respect to any trait X can be formally expressed by the condition

$$\frac{dR}{dX} = 0 \,. \tag{1.13}$$

If trait X is independent of age and affects both survival, $l(a, X)$, and reproduction, $m(a, X)$, at various ages, then together with (1.12) this condition yields

$$\int_0^\infty \left(\frac{\partial l(a,X)}{\partial X} \, m(a,X) + \frac{\partial m(a,X)}{\partial X} \, l(a,X) \right) \, da = 0 \,. \tag{1.14}$$

Extracting the product $l(a, X)\, m(a, X)$ and using the shorthand notation

$$\frac{\frac{\partial l(a,X)}{\partial X}}{l(a,X)} \equiv \acute{l}_X(a,X) \tag{1.15}$$

for the relative change in survival with respect to trait X and an analogous notation for the relative change in reproduction, the condition can be expressed as

$$\int_0^\infty \left(\acute{l}_X(a,X) + \acute{m}_X(a,X) \right) \, l(a,X)\, m(a,X) \, da = 0 \,. \tag{1.16}$$

Finally, note that dividing by the life-time reproduction given in (1.12) yields the average value (indicated by the bar) of the relative change (indicated by the acute accent) in survival,

$$\frac{\int_0^\infty \acute{l}_X(a,X)\, l(a,X)\, m(a,X) \, da}{\int_0^\infty l(a,X)\, m(a,X) \, da} \equiv \bar{\acute{l}}(a,X) \,, \tag{1.17}$$

and analogously for reproduction. Consequently, Condition (1.13) is equivalent to

$$\bar{\acute{l}}_X(a,X) + \bar{\acute{m}}_X(a,X) = 0 \,. \tag{1.18}$$

The value of X that maximizes fitness corresponds to the point where the average relative change in survival plus the average relative change in reproduction with respect to trait X equals zero.

If trait $X(a)$ only affects survival and reproduction at a specific age a, i.e. $l(x, X(a))$ and $m(x, X(a))$, then (1.14) reduces to

$$\frac{d\,\mu(a,\,X(a))}{dX(a)}\,v(a) \;=\; \frac{d\,m(a,\,X(a))}{dX(a)}\,. \tag{1.19}$$

The value of $X(a)$ that maximizes fitness corresponds to the value where the change in mortality $\mu(a, X(a))$ with respect to trait $X(a)$ at age a times the reproductive value $v(a)$ at age a equals the change in reproduction $m(a, X(a))$ with respect to the trait at age a.

There are alternative ways to find the optimal schedule for a trait. Being optimal implies achieving the best life history strategy over the entire lifespan, which is equivalent to doing this at every age. Since the future does not influence the past, the optimal strategy at every age is to maximize

$$\begin{gathered}\textit{current reproduction } + \\ \textit{survival to next age } \cdot \textit{ remaining reproduction}\end{gathered} \tag{1.20}$$

assuming the individual is alive at that age. Maximizing this quantity is equivalent to maximizing the current reproductive value given by (1.1), which can be seen using the discrete-time formulation

$$v_a \;=\; \frac{e^{r\,a}}{l_a} \sum_{i=a}^{\infty} e^{-r\,i}\, l_i\, m_i\,. \tag{1.21}$$

Extracting the first term from the sum yields

$$v_a \;=\; m_a \;+\; \frac{e^{r\,a}}{l_a} \sum_{i=a+1}^{\infty} e^{-r\,i}\, l_i\, m_i\,.$$

Multiplying the sum by a factor of $1 = l_{a+1}e^{r} \,/\, l_{a+1}\,e^{r}$ and letting $p(a)$ be the probability of surviving from age a to $a{+}1$, $p(a) \;=\; l(a{+}1)\,/\,l(a)$, the nature of the general life history trade-off becomes apparent:

$$v_a \;=\; m_a \,+\, p(a)\,e^{-r}\,v_{a+1}\,. \tag{1.22}$$

The first term captures the profits obtained from current reproduction, m_a. The second term captures the future prospects. The future prospects depend on the chance of getting there, i.e. surviving the age interval ($p(a)$, discounted by population growth e^{-r}) and future reproductive potential, which is reproductive value v_{a+1} at the next age (see [27, Chap. 5] for review).

Current reproduction trades off with future survival and reproduction. On the one hand, this trade-off could be due to a direct negative effect of reproduction on survival. Mating activities, for instance, could be risky. Also, reproduction could cause damage that negatively affects future breeding attempts. Whereas this direct negative effect is not necessarily observed in all species, a negative indirect link becomes apparent if survival and reproduction are understood as distinct processes that compete for limited resources.

Schaffer [176] stated that the *general life-history problem* is to allocate restricted resources between survival and reproduction in a way that maximizes an individual's fitness. To approach this problem Williams [213] introduced the reproductive effort model, where reproductive effort is defined as the fraction of energy devoted to reproduction. Williams [213] conjectured that, at every age, resources are allocated to maximize the remaining reproductive contribution of an individual that already survived to that age, i.e. the reproductive value. From Bellman's principle (see [12] and Sect. 4.3 of this manuscript) we know that maximizing reproductive value at every age is equivalent to maximizing reproductive value at age zero. In that way Williams [213] anticipated Taylor et al.'s [192] result that "*[m]aximizing the reproductive value at age zero is mathematically equivalent to maximizing the ultimate rate of increase*". Extensive treatments of the evolution of optimal life histories can be found in [186] and [169].

I want to emphasize how reproductive value emerges again and again as an important quantity. Not only was it proposed as a measure of senescence [149] – it was also proved to be a measure of fitness [192] and a central quantity for solving the general life-history problem [213].

1.2.6 Interesting Recent Developments

In Chaps. 4 and 5, I will develop models to explain the evolution of senescence that focus on the age-patterns of mortality, fertility and growth using the concepts outlined above. Reproductive-effort models were developed in the 1970s to understand when iteroparity (repeated breeding) is favored over semelparity (single breeding event, in which reproduction is fatal) (see [62], [175] and [31]). The shape of the age-trajectory of mortality itself attracted little interest. Instead, mortality was assumed to follow a particular pattern, for example to be constant, to be stepwise constant (distinguishing only between a juvenile and an adult period) or to follow an exponential pattern.

Some recent models of the evolution of senescence, however, do focus on the age-trajectory of mortality in conjunction with age-trajectories of growth, reproduction and transfers. These models draw heavily on the concept of allocation of restricted resources and on dynamic optimization techniques (see [12] and Sect. 4.3).

Abrams and Ludwig [5] develop a theoretical model based on the disposable soma theory [97] and find that many different mortality trajectories can be optimal, an exponential increase being only one possible outcome. The model, however, does not allow for a decline in mortality with age.

Mangel and Bonsall [120] also show that a diversity of optimal mortality trajectories is possible when mortality is viewed as a result of multiple physiological processes as well as when mortality is the consequence of growth and metabolism and associated damage. In their model, mortality can decrease over some ages before it ultimately increases. Another recent model by Mangel and Munch [121] that focuses on compensatory growth derives mortality as result of growth and damage. The approach taken by Mangel and colleagues shows that optimal age-patterns of mortality can decrease if mortality is, at least in part, determined by physiological state. They point out the importance of "reunifying the connections between the biology of aging and demography" [120, p. 357]. Munch and Mangel [131] recently showed that mortality can follow various patterns at juvenile ages.

Dynamic programming models that optimize resource allocation to growth, reproduction and repair of somatic damage based on the disposable soma theory of aging have been studied intensively by Kozlowski and Cichon [37, 38, 39, 102, 103]. Their models do not allow mortality to decline with age. Drenos and Kirkwood [52] also describe a mathematical model based on the disposable soma theory. In their model the optimal level of investment in repair is always less than that required for non-senescence.

An approach that explicitly questions when senescence can be escaped is given by Gardner and Mangel [64]. They develop a stage-based model and find that the strength of selection can, under some circumstances, increase with age for clonal organisms.

Travis [196] claims that, in a spatially structured population, a determinate lifespan can evolve with an optimal specific age of death, but in a freely mixing population with global dispersal evolution selects for individuals with ever-increasing lifespan. In a working paper, Doncaster and Seymour [50] demonstrate that ever-extending reproductive life can be optimal in populations with density regulated recruitment, e.g., in the case of Bristlecone Pines. If seeds can be established only on a patch freed by the death of an adult, it pays to outlive your neighbors to ensure that your offspring can occupy the newly opened space.

Sozou and Seymour [183] show that mortality does not necessarily have to increase, i.e. that non-senescence can be locally optimal, if the potential onset of deterioration is sufficiently rapid or early. Interestingly, they find that "for all forms of profile considered, conditions can be found for which a strategy involving no ageing is locally optimal".

In a recent paper, Chu and Lee [36] study the conditions under which transfers from adult to offspring can be optimal. Applying dynamic optimization techniques and the idea of optimal resource allocation, they model the co-evolution of survival and transfers. A recent working paper by Robson and Kaplan [168] derive a dynamic optimization model for the evolution of the human mortality pattern incorporating investment in quantity and quality of somatic capital and a budget constraint that reflects intergenerational transfers. These models can explain why mortality declines during development and why evolution licences a substantial period of post-reproductive life in humans.

With the models I am going to develop, I will not be focusing on a single species such as humans. I wish to understand more generally under what conditions what pattern of mortality can be expected. In particular, I want to study if and when non-senescence can be optimal. My work is the first systematic attempt to find the characteristics that determine when senescence is optimal and when it is not. I will not focus on lifespan. A species with a short lifespan can still have a non-senescent life history. The length of life only reflects different time scales that different species live on. This would be a different question: When is it optimal to live on what time scale? Instead I ask: When is it optimal to live under what qualitative mortality pattern?

My modeling strategy is to exploit the power of focused simplicity. The models will be kept as simple as possible, including only necessary ingredients that are chosen based on my particular question.

1.3 Orientation

In the following two chapters I discuss Hamilton's paper on the molding of senescence [75], disproving his dogmatic claim that senescence is inevitable and pointing out deficiencies of Hamilton's framework. Given the theoretical issues and empirical evidence, I come to the conclusion that life histories are likely to be shaped largely by optimization rather than by a burden of deleterious mutations, at least over ages where the bulk of life-time reproduction is realized.

In the subsequent two chapters, I develop optimization models to determine the optimal pattern of survival and reproduction over the life course of a species. The models in Chap. 4 are based on the state-variable size. The Chapter makes the case for negative senescence, i.e. the models show that, theoretically, senescence is not an inherent part of life. The model in Chap. 5 is built around the state-variable "vitality" and takes into account and addresses some of the deficiencies of the size-based models. The vitality model demonstrates that the space of optimal life histories is wide and covers a broad range of senescent and non-senescent strategies.

The final chapter, Chap. 6, emphasizes the need to connect the world of mutation accumulation and the world of optimization. I also suggest directions for future research on the evolution of senescence.

Part I

Hamilton

2

Hamilton's Indicators of the Force of Selection

2.1 Introduction

To quantify the force of selection, Hamilton derived expressions for the change in fitness with respect to age-specific mutations. Hamilton's indicators are decreasing functions of age. He concluded that senescence is inevitable: survival and fertility must decline with age. I show that an alternative parametrization of mutational effects leads to indicators that can increase with age. I then consider the case of deleterious mutations with age-specific effects. In this case, it is the balance between mutation and selection pressure that determines the equilibrium number of mutations in a population. In this balance the effects of different parameterizations cancel out, but only to a linear approximation. I show that mutation accumulation has little impact at ages when this linear approximation holds. When mutation accumulation matters, nonlinear effects become important and the parameterizations of mutational effects make a difference. The results also suggest that mutation accumulation may be relatively unimportant over most of the reproductive lifespan of any species.

Senescence can be defined as an increase in mortality and/or a decrease in fertility with age. Is senescence a universal characteristic of life? It is not obvious from an evolutionary perspective why it should be. Early in life, when individuals develop and grow, mortality falls and reproductive potential increases. Why is it that these age-patterns cannot persist, in some form, with mortality continuing to decline and reproductive capacity continuing to increase? George C. Williams [212, p. 398] wrote:

"It is indeed remarkable that after a seemingly miraculous feat of morphogenesis a complex metazoan should be unable to perform the much simpler task of merely maintaining what is already formed".

William D. Hamilton's influential article on "The Moulding of Senescence by Natural Selection" [75, 76] provides a reason why senescence "cannot be avoided by any conceivable organism". Hamilton combines insights about the evolution of senescence [126, 212] with concepts and models of population dynamics [115]. Hamilton asserts that " senescence is an inevitable outcome of evolution". Did Hamilton genuinely prove that senescence is theoretically inevitable?

2.2 Hamilton's Derivations

How does a mutation that acts only at a specific age a influence the evolutionary success of an individual? Does it matter if this age is early or late in life? Hamilton [75] built on the insight of Medawar [126] that later-acting genes should be under weaker selection than earlier-acting ones due to the unavoidable decline in the number of survivors at higher and higher ages. A genetically-determined fatal disease that struck only at post-reproductive ages would be entirely out of reach of the force of selection.

2.2.1 The Framework

To quantify the force of selection Hamilton considered age-specific, mutation-induced changes in fitness. Hamilton used the most widely-accepted measure of Darwinian fitness, the intrinsic rate of population increase r, implicitly defined by the discrete version of the Lotka equation

$$\sum_{x=0}^{\infty} e^{-r\,x}\, l_x\, m_x \;=\; 1\,. \tag{2.1}$$

The function l_x gives the chance of survival to age x. The function m_x measures the amount of reproduction at that age. If the population is stable, as assumed by Hamilton, then each combination of an age-specific maternity function m_x and an age-specific survival function l_x is associated with exactly one real r that satisfies (2.1).

The survival function l_x is defined as the product of the probabilities p_a of survival from age a to $a+1$:

$$l_x = p_0\, p_1 \,\ldots\, p_{x-1} \,, \tag{2.2}$$

with

$$l_0 = 1 \,.$$

The age-specific survival probabilities p_a depend on the instantaneous death rate μ_t, the force of mortality between age a and $a+1$, via

$$p_a = e^{-\int_a^{a+1} \mu_t\, dt} = e^{-\bar{\mu}_a} \,. \tag{2.3}$$

The cumulative mortality in the exponent reflects the average mortality during that time interval, denoted by $\bar{\mu}_a$.

2.2.2 Hamilton's Indicator of Survival

By taking the derivative of (2.1) with respect to $\ln p_a$ and rearranging, Hamilton derived his basic result:

$$H^\dagger \equiv \frac{dr}{d \ln p_a} = \frac{\sum_{x=a+1}^{\infty} e^{-r\,x}\, l_x\, m_x}{\sum_{x=0}^{\infty} x\, e^{-r\,x}\, l_x\, m_x} \,. \tag{2.4a}$$

Note that (2.3) implies that $H^\dagger$ can also be expressed as:

$$H^\dagger \equiv -\frac{dr}{d\bar{\mu}_a} \,. \tag{2.4b}$$

The value of $H^\dagger$ is a measure of the force of selection. It captures the change in fitness r induced by an increase in $\ln p_a$. An increase in $\ln p_a$ is equivalent to a reduction in average mortality $\bar{\mu}_a$ between age a and $a+1$. This sensitivity of fitness to changes in age-specific survival is captured by the ratio of remaining reproduction, the numerator in (2.4a), to generation time, the denominator. Because $H^\dagger$ declines as age increases, Hamilton concluded that the force of selection must decline with age.

2.3 Alternative Indicators

2.3.1 Different Parametrization

Hamilton's conclusion hinges on the particular parametrization he chose for the nature of the effect of a mutation. Equally reasonable, alternative forms would have been dr/dp_a, dr/dq_a, $dr/d\ln q_a$ or $dr/d\ln\bar{\mu}_a$, where q_a is the probability of dying ($q_a = 1 - p_a$) and $\bar{\mu}_a$, as noted above, equals $-\ln p_a$. The results are as follows:

$$\frac{d\,r}{d\,p_a} = \frac{1}{p_a}\,H^{\dagger}, \tag{2.5a}$$

$$\frac{d\,r}{d\,q_a} = -\frac{1}{p_a}\,H^{\dagger}, \tag{2.5b}$$

$$\frac{d\,r}{d\,\ln q_a} = -\frac{q_a}{p_a}\,H^{\dagger} \tag{2.5c}$$

and

$$\frac{d\,r}{d\,\ln\bar{\mu}_a} = -\,\bar{\mu}_a\,H^{\dagger}\,. \tag{2.5d}$$

Strikingly, the expressions in (2.5a-d) can increase in absolute value with age – in contrast to $H^{\dagger}$, which always declines.

2.3.2 When Selection Pressure Increases

Consider, for instance, (2.5d). At pre-reproductive ages the value of $dr/d\ln\bar{\mu}_a$ is entirely determined by $\bar{\mu}_a$, as $H^{\dagger}$ is constant before maturity. At reproductive ages the change in fitness with respect to mortality increases from age a to $a+1$ if

$$\left|\frac{d\,r}{d\,\ln\bar{\mu}_a}\right| < \left|\frac{d\,r}{d\,\ln\bar{\mu}_{a+1}}\right|.$$

Substituting (2.5d) and (2.4a), and using the notion of reproductive value,

$$v_a = \frac{e^{r\,a}}{l_a}\sum_{x=a}^{\infty} e^{-r\,x}\,l_x\,m_x\,, \tag{2.6}$$

this inequality can be rearranged to give the following condition,

$$\left(\frac{\bar{\mu}_{a+1} - \bar{\mu}_a}{\bar{\mu}_{a+1}}\right)\frac{v_{a+1}}{m_{a+1}} > 1\,. \tag{2.7}$$

Hence, the value of $dr/d\ln\bar{\mu}_a$ will increase with age if $\bar{\mu}_a < \bar{\mu}_{a+1}$ and if future reproductive value is sufficiently large compared to fertility m_{a+1}. Taking into account the fact that (2.1) must hold, the inequality in (2.7) can be rearranged as

$$\left(\frac{\bar{\mu}_{a+1} - \bar{\mu}_a}{\bar{\mu}_{a+1}}\right) \frac{e^{r(a+1)}}{l_{a+1}} \left(1 - \sum_{x=0}^{a} e^{-r\,x}\, l_x\, m_x\right) > m_{a+1}\,. \tag{2.8}$$

This inequality determines trajectories for m_{a+1} that lead to increasing sensitivity of fitness to changes in mortality over age given a specified, increasing path for $\bar{\mu}_a$. The survival and fertility functions plotted in Fig. 2.1 and the resulting indicators $dr/d\ln\bar{\mu}_a$ and $dr/d\ln p_a$ plotted in Fig. 2.2 provide an illustrative example.

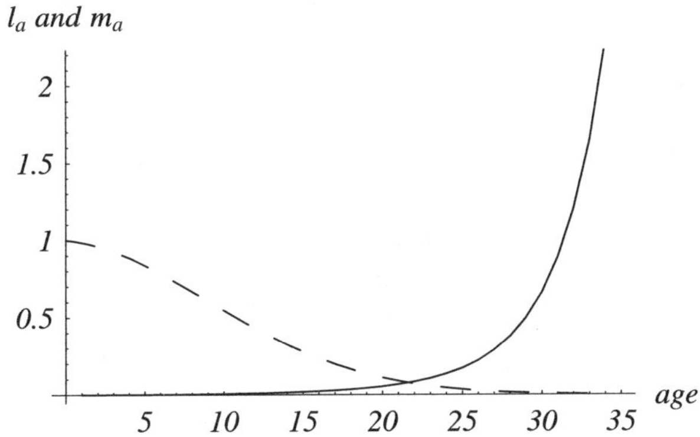

Fig. 2.1. Example of survival and maternity functions l_a and m_a (If age-specific survival probabilities p_a change according to $p_a = p_0^a$ with $p_0 < 1$, then the average force of mortality between age a and $a+1$ is given by $\bar{\mu}_a = -\ln p_0^a = -a\ln p_0$. Maternity m_{a+1} was chosen to be 0.01 units smaller than the left-hand side of the inequality in (2.8), setting $r = 0$, $p_0 = 0.99$ and $m_0 = 0$. By age 34, survival falls to 0.25%. After age 34, I fixed age-specific survival p_a at its level of $p_{35} = 0.70$ corresponding to $\bar{\mu}_{35} = 0.35$ and adjusted m_a to a constant level of 133.265 such that (2.1) is fulfilled.)

2.3.3 Fertility Indicators

The quantity Hamilton derived for the force of selection on age-specific mutations that affect fertility is

$$H^* \equiv \frac{d\,r}{d\,m_a} = \frac{e^{-r\,a}\, l_a}{\sum_{x=0}^{\infty} x\, e^{-r\,x}\, l_x\, m_x}\,. \tag{2.9}$$

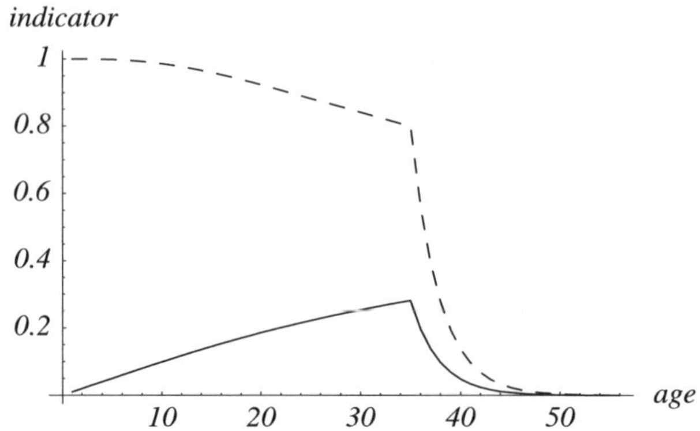

Fig. 2.2. Comparison of $H^\dagger = \frac{dr}{d\ln p_a}$ (*dashed line*) with $\frac{dr}{d\ln \bar{\mu}_a}$ (*solid line*) (While Hamilton's indicator $H^\dagger$ declines, the alternative one increases until age 34. The increase would have continued if m_{a+1} had been further determined by the inequality in (2.8). This, however, would result in a trajectory for m_a that would rise to enormous heights. Also note that Hamilton's indicator is greater than the alternative indicator, especially before age 35. This implies a considerably stronger force of selection on age-specific mutations that affect mortality.)

Hamilton considered survival effects on a log scale: He could have done the same for reproduction, calculating

$$\frac{dr}{d\ln m_a} = m_a H^* . \tag{2.10}$$

Hamilton's indicator in (2.9) necessarily declines with age but the alternative indicator in (2.10) can increase with age, depending on the trajectory of m_a.

Table 2.1 summarizes the direction of changes over age of the various indicators of the force of selection. The differences in the dynamics are due to the nonlinearity of logarithmic and exponential transformations.

2.3.4 Are Some Indicators Better?

Charlesworth [27, p.191], who reconstructed Hamilton's results, suggested that "genetic effects on survival probabilities are more likely to be additive on a log scale." His conjecture implies that mutations have additive effects on mortality. Indeed, both of Hamilton's indicators $H^\dagger = -dr/d\bar{\mu}$ and $H^* = dr/dm$ can be interpreted as assuming that mutations additively affect average mortality $\bar{\mu}$ and fertility m. This

Table 2.1. Various indicators of the force of selection in Hamilton's framework

Indicator	Change with age a
$\frac{d\,r}{d\,\ln p_a}$	$-$
$\frac{d\,r}{d\,p_a}$	$+$ or $-$*
$\frac{d\,r}{d\,q_a}$	$+$ or $-$
$\frac{d\,r}{d\,\ln q_a}$	$+$ or $-$
$\frac{d\,r}{d\,\ln \bar{\mu}_a}$	$+$ or $-$
$\frac{d\,r}{d\,m_a}$	$-$
$\frac{d\,r}{d\,\ln m_a}$	$+$ or $-$

* "+ or −" means that the change with age can be positive or negative, depending on the trajectories of m_x and l_x.

is plausible because additive risk models are widely used, most commonly in evolutionary modeling [23, 29]. The indicators $\bar{\mu}H^{\dagger}$ and mH^{*} capture the effect of a proportional change in $\bar{\mu}$ and m. Proportional-hazard models in general and Cox proportional-hazard models [45] in particular are frequently used in demographic and epidemiological research.

Deleterious mutations influence the internal condition of an organism. Internal conditions are known to interact with the environment [163, 214]. These interactions affect mortality in a non-additive manner. The idea that traits are likely to combine non-additively is also supported by recent work by Promislow [160] and Spencer and Promislow [184] which concerns the network structure of genes and epistasis respectively.

Whether age-specific mutations act proportionally or additively has been a question for empirical research. Support for the preeminence of proportional hazards comes from *Drosophila.* The study by Promislow and colleagues [161] of additive genetic variance favors proportional hazards. In the papers by Good and Tatar [68] and Mair et al. [116] change in current nutrient conditions affects mortality in a proportional manner. Furthermore, many mutants extend lifespan in *Drosophila* because they reduce mortality proportionally [87, 112, 170]). An exception is the work on the mutant chico [197]. Evidence for proportional hazards also comes from baboons [14] and mice [60] [1].

Numerous demographic and epidemiological analyses of risk factors have found that proportional effects are more common than additive effects. In particular, the impact of genetic polymorphisms, such as ApoE 2, 3 and 4, on mortality has been modeled by proportional hazards [66]. Empirical results reviewed by Promislow and Tatar [158] support the proportional-hazard assumption, suggesting that mutations act additively on log-mortality rather than log-survival. Hence, it seems plausible that the indicators $\bar{\mu}H^{\dagger}$ and mH^{*} will prove at least as valid as Hamilton's indicators.

2.3.5 Optimization vs. Mutational Burden

How mutations affect fitness is the focus of a vast literature [17, 27, 46, 54, 73, 74, 96]. Since Medawar [126] and Hamilton [75], many biologists have considered the sensitivity of fitness with respect to age-specific changes in survival or fertility [23] as an indicator of selection pressure. A key issue is whether age-patterns of mortality and fertility are molded by adaptive optimization processes or by the burden of non-adaptive mutations [2, 27, 147, 148]. Note that, in either case, an increase in mortality or a decrease in fertility is a byproduct of evolutionary processes. In the former case, senescence can arise as a side effect of an optimal balance between linked traits that effect fitness, and in the latter case senescence emerges as the weakening selection pressure is less and less successful in eradicating deleterious mutations.

Optimization models can be solved without using Hamilton's indicators [200]. If the age-patterns mainly reflect the age-specific burden of mutations, then Hamilton's indicators are not sufficient. Age-specific levels of birth and death rates depend not only on the selection pressure but also on mutation rates. In the following section I analyze this balance.

2.4 Mutation–Selection Balance

How do the alternatives of parametrization in Table 2.1 affect the equilibrium number of deleterious mutations at each age? In particular,

[1] I thank Marc Tatar for emphasizing the preeminence of proportional hazards and for pointing me to the relevant empirical evidence.

is the magnitude of mutation accumulation great enough to mold the trajectory of mortality?

The equilibrium number of mutations under mutation–selection balance can be approximated by the ratio of the total mutation rate ν (i.e., the hazard of a mutation from a set of possible mutations) and the change in fitness r:

$$\bar{n} \approx \frac{\nu}{\frac{dr}{dn}}, \tag{2.11}$$

where n denotes the number of mutations and $\bar{n}$ denotes the equilibrium number [27, pp. 125-126]. The approximation holds if ν and $\bar{n}$ are small. Using the chain rule, the derivative in (2.11) can be factored into the change in fitness with respect to survival or fertility and the effect on survival or fertility of having n mutations:

$$\frac{dr}{dn} = \frac{dr}{df}\frac{df}{dn}, \tag{2.12}$$

where f could be any of the denominators in Table 2.1.

2.4.1 Additive vs. Proportional Parametrization

Consider a mutation that has a small effect δ on mortality. Then f is equivalent to

$$\mu_a(n) = \mu_a(0) + n\,\delta \tag{2.13a}$$

in the additive case and

$$\ln \mu_a(n) = \ln \mu_a(0) + n\,\delta \tag{2.13b}$$

in the proportional case. From (2.11), (2.12) and Table 2.1 it follows that

$$\bar{n} \approx \frac{\nu}{h_a^\dagger\,\delta} \tag{2.14a}$$

in the additive case and

$$\bar{n} \approx \frac{\nu}{\mu_a(0)\,h_a^\dagger\,\delta} \tag{2.14b}$$

in the proportional case. In these ratios $h_a^\dagger$ denotes remaining reproduction at age a of an individual with no deleterious mutations. It is related to Hamilton's indicator via $h_a^\dagger = H_a^\dagger T$, where T captures generation time.

Combining (2.13) and (2.14) leads to the result

$$\mu_a(\bar{n}) \approx \mu_a(0) + \frac{\nu}{h_a^\dagger} \tag{2.15a}$$

in the additive case and

$$\mu_a(\bar{n}) \approx \mu_a(0) \exp\left(\frac{\nu}{\mu_a(0)\, h_a^\dagger}\right) \tag{2.15b}$$

in the proportional case. If mutations are rare, i.e. if $\nu/\mu_a(0)$ is small, then the formula for the proportional case can be approximated by

$$\mu_a(\bar{n}) \approx \mu_a(0)\left(1 + \frac{\nu}{\mu_a(0)\, h_a^\dagger}\right) = \mu_a(0) + \frac{\nu}{h_a^\dagger}\,. \tag{2.16}$$

Hence, if ν and $\bar{n}$ are small enough that the approximations in (2.11) and (2.16) hold, then mutation accumulation will result in about the same age-specific mortality regardless of whether mutations have additive or proportional effects.

2.4.2 A Simple Box Model

If $\bar{n}$ is large, an alternative approach is necessary. Several helpful models have been developed (e.g. [95, 127, 128, 142]); for a review see [17, 27]. A recent general model by Steinsaltz, Evans, and Wachter [187] includes earlier models as special cases.

A solution based on a simple box model similar to that of Kimura and Maruyama [95] can be readily developed. Assume a haploid, asexual population that is stationary in size. Further assume that mutations affect only one age class, to ensure that the equilibrium numbers of mutations are independent across ages. Focus on a single age a. Individuals are sorted into boxes according to their number of mutations at age a. Let $N(n)$ be the number of individuals in box n and let N be the total, constant population size at age a. In mutation–selection balance, the proportions $N(n)/N$ are fixed. Denote the lifetime reproduction of an individual in box n by $R(n)$. Let ν be the probability of passing on a new, additional mutation to the next generation. Assume that mutations occur successively, i.e. it is not possible to jump over boxes. Ignore back mutations. Mutations are deleterious, therefore $R(0) > R(1) > R(2)... > R(K)$, K being some maximum number.

The number of individuals $N(n)$ in box n is given by the inflow of individuals minus the outflow per generation,

$$N(n) = N(n-1)\,R(n-1)\,\nu + N(n)\,R(n)\,(1-\nu)\,. \tag{2.17}$$

It follows immediately that reproduction in box zero is

$$R(0) = \frac{1}{1-\nu}\,. \tag{2.18}$$

In the case of mutations that affect mortality, the lifetime reproduction of individuals in the n'th box is given by

$$R(n) = \sum_{x=0}^{a-1} l_x\, m_x + e^{\mu_a(0) - \mu_a(n)} \sum_{x=a}^{\infty} l_x\, m_x\,. \tag{2.19}$$

This result can be expressed as

$$R(n) = R(0) - \Delta(n)\, h_a^{\dagger}\,, \tag{2.20}$$

where $\Delta(n)$ is the fraction of remaining reproduction $h_a^{\dagger}$ that is lost due to carrying n mutations. In the additive case

$$\Delta(n) = 1 - e^{-\delta\, n} \tag{2.21a}$$

and in the proportional case

$$\Delta(n) = 1 - e^{-\mu_a(0)\,(\exp[\delta\, n] - 1)}\,. \tag{2.21b}$$

It follows from (2.17) and (2.20) that

$$N(n) = \frac{N(0)}{\prod_{k=1}^{n} \Delta(k)}\, R(0)^{n+1} \left(\frac{\nu}{h_a^{\dagger}}\right)^{n} \prod_{k=1}^{n-1} \left(R(0) - \Delta(k)\, h_a^{\dagger} \right). \tag{2.22}$$

The equilibrium number of mutations is the average over all boxes, i.e.

$$\bar{n} = \frac{\sum_{n=0}^{K} n\, N(n)}{\sum_{n=0}^{K} N(n)}\,. \tag{2.23}$$

Figure 2.3 plots the equilibrium number of mutations over age in the additive versus proportional case for the example presented in Fig. 2.1 and 2.2. As a second example I consider female mortality, as given in the Swedish life table for 1778-82. Results are shown in Figs. 2.4 and 2.5.

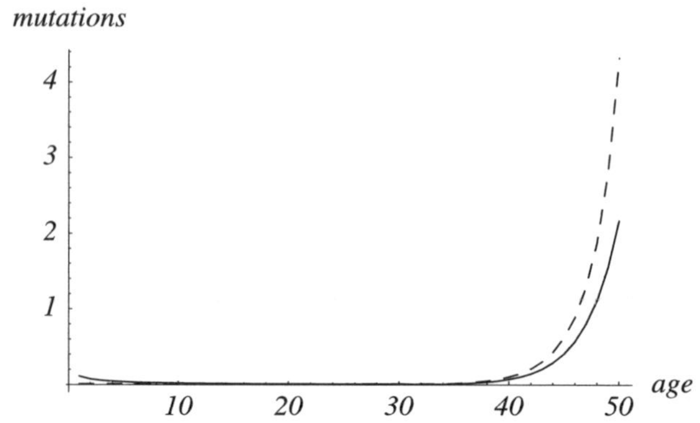

Fig. 2.3. Equilibrium number of mutations: additive (*dashed*), proportional (*solid*) (I assume that mutation pressure $\nu = 0.001$. Furthermore, I assume that a mutation at any age reduces remaining reproduction by about ten percent in both the additive and proportional case. This refers to an average reduction in the proportional case since $\Delta(n)$ depends on the level of mortality at age a, as can be seen from (2.21b). Specifically, $\delta = 0.1$ in (2.21a) and $\delta = 0.35$ in (2.21b). While in the Hamiltonian case of an additive hazard the number of mutations remains low and then increases with age, proportional effects lead to an age-specific mutational load that declines at young ages. In the example only one quarter of one percent of individuals are alive at age 34. Before this age the mutational load is close to zero. After this age, however, the equilibrium number of mutations rises sharply.)

The values of $h^\dagger$ that determine the number of mutations in Figs. 2.3 and 2.4 are calculated using specific *initial* fertility and mortality schedules. The mutations, however, will raise mortality, producing a new schedule that determines a new $h^\dagger$, as illustrated in Fig. 2.5. These dynamics are beyond the scope of this chapter. Note, however, that higher hazard rates would reduce the fitness costs of a change in age-specific mortality. Thus, more mutations would accumulate and the difference between additive and proportional parameterizations would be larger than predicted by my conservative estimate. A general treatment that takes into account interactions between ages is given by Steinsaltz, Evans, and Wachter [187].[2]

[2] I thank Kenneth W. Wachter and Brian Charlesworth for helping me considerably with this section.

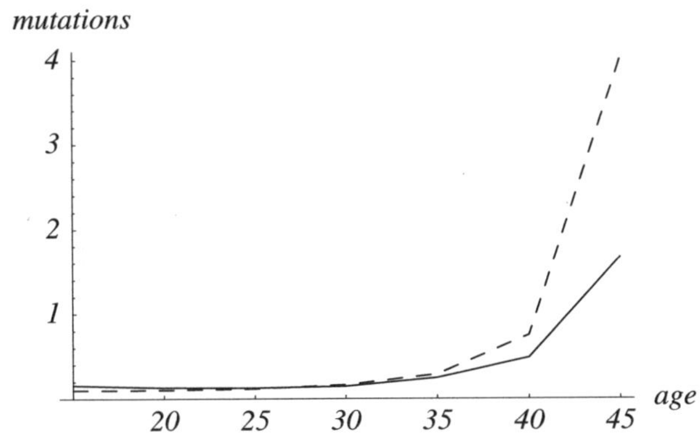

Fig. 2.4. Equilibrium number of mutations: additive (*dashed*), proportional (*solid*) (The example is based on female mortality as given in the Swedish life table for 1778-82, for seven 5-year age-groups, beginning at age 15. Since the Swedish population was growing at that time, I normalized reproduction to ensure $R = 1.00$. I consider a deleterious mutation that reduces remaining reproduction at any age by about one percent, either in an additive or in a proportional way, i.e. $\delta = 0.01$ in (2.21a) and $\delta = 0.7$ in (2.21b), and I assume a mutation pressure of $\nu = 0.001$. The difference between the additive and proportional case increases at higher ages, as levels of remaining reproduction decline. A slight decrease in the equilibrium number of mutations from the first to the second age-group can be observed.)

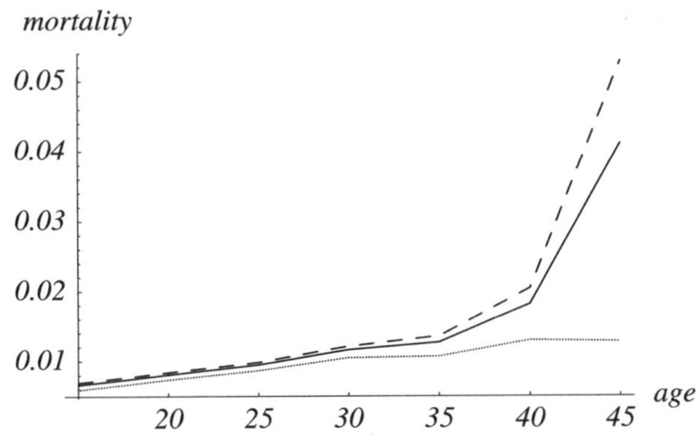

Fig. 2.5. Mortality: additive (*dashed*), proportional (*solid*), initial mortality $\mu_a(0)$(*dotted*) (Initial mortality is from the Swedish life table for 1778-82, females, for seven 5-year age-groups, beginning at age 15.)

2.5 The Importance of Mutation Accumulation

The age-trajectory of mortality can be decomposed into three parts: one component is due to the accumulation of unfavorable mutations, another fraction results from selection processes that optimize the trade-offs necessitated by resource limitations, and the remaining fraction can be attributed to unavoidable, external risks of death. How strong is the influence of mutation accumulation?

The relative impact of mutation accumulation on the molding of the mortality trajectory is crucially determined by the ratio of mutation pressure ν to remaining reproduction $h_a^\dagger$, as indicated by (2.14). The larger the value of ν, the more influential is mutation accumulation. But what is the magnitude of ν? Keightley and Charlesworth [92] point out that the rate of deleterious mutations per haploid genome in *C. elegans* in protein coding genes is about 0.5 per generation. Kimura and Maruyama [95] and Drake et al. [51] suggest mutation rates per genome per generation of about 0.1 and between $0.1 - 100$, respectively. More recent publications estimate the genomic rate of deleterious mutations in humans to be at least 1.6 [55] or even 3 [133] per generation.

If the fraction of mutations that exclusively affect mortality at a specific age is low, then these values could be consistent with a value of $\nu = 0.001$. If ν is 0.001, then Fig. 2.6 suggests that the influence of mutation accumulation is likely to be small over the major part of reproductive life. This remains speculation, however, until the magnitude of ν is estimated empirically. Abrams [2] provides suggestive evidence that the importance of mutation accumulation is likely to be small relative to the importance of optimization among trade-offs. Partridge [147] points out that little evidence can be found in favor of mutation accumulation but considerable evidence can be found to confirm the importance of trade-offs.

The conclusions drawn above and in the previous section were reached on the basis of a specific model of mutation accumulation. In general cases covered by the solutions given by Steinsaltz, Evans, and Wachter [187], the form of the mutation–selection equilibrium depends on the extent of assumed genetic recombination. At both extremes, in the absence of recombination (Equation 9 in their article) and in the presence of free recombination (Equation 27), the parametrization of the mutational effect, i.e. whether the effect is additive or proportional, influences the mutation–selection equilibrium.

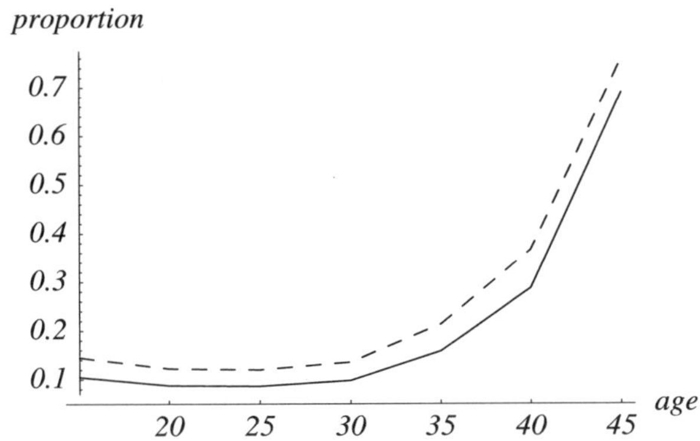

Fig. 2.6. Proportion of mortality explained by mutation accumulation: additive (*dashed*) vs. proportional (*solid*) case (The fraction $1 - \mu_a(0)/\mu_a(\bar{n})$ indicates the proportion of equilibrium mortality that can be explained by the accumulation of mutations. For the example of Swedish females, when $\nu = 0.001$, mutation accumulation explains less than a third of total mortality. At ages 45-50, however, when mortality is high and fertility is low, mutation accumulation accounts for the bulk of total mortality. Note that this illustrative example does not pertain to actual Swedish mortality but to the hypothetical outcome of one round of mutation accumulation: see Sect. 3.1 for further discussion.)

2.6 Conclusion

Hamilton stated that the force of selection inevitably has to decline with age, even "in the farthest reaches of almost any bizarre universe" [76]. He concluded that the declining selection pressure would mold the age-pattern of mortality in a way that mortality is lowest at reproductive maturity and "trails upward indefinitely at the right ... roughly asymptotic to the age of the ending of reproduction" [76, p. 119]. Hamilton's claim about the inevitability of senescence has been generally accepted, but it can be disproved, even adopting his restrictive assumptions. As shown above, alternative indicators can be derived, within Hamilton's own framework, that can result, in some circumstances and over some age ranges, in an increasing force of selection with age, thus contradicting the basis for his claim.

The results of this chapter strengthen the view that demographic schedules of mortality and fertility appear to be shaped largely by optimization of trade-offs rather than by mutation accumulation. Only at ages when remaining reproduction is low does the influence of mutation accumulation appear to become predominant. At those ages, different parameterizations lead to different conclusions about the equilibrium number of mutations.

Some important empirical research questions are suggested by the theoretical findings of this chapter. Does the age-specific mutation rate ν change with age? If so, what is the age-trajectory of ν?

3

Further Challenges

Hamilton's claim of the inevitability of senescence can be disproved even within his own framework. Furthermore, his framework has several limitations. In this chapter theoretical and empirical issues that weaken his approach as the main explanation for the evolution of senescence will be discussed. Building on Medawar [126] and Williams [212], Hamilton wrote the pioneering *first* chapter on the moulding of senescence.

I draw two main conclusions.

- First, Hamilton's basic notion – that the age-pattern of mortality is an inverse function of the age-pattern of his indicator – is wrong. For both his indicator and the other indicators in Table 2.1 the relationship between the indicator and mortality is so complicated that sophisticated modeling is required.
- Second, several theoretical arguments as well as the bulk of empirical findings suggest that mutation accumulation is of secondary importance in molding the age-trajectories of mortality across the varied species of life. The primary force appears to be adaptation, i.e. the concept that patterns of aging are a byproduct of optimization of trade-offs. Hence, deep understanding of the evolution of aging requires optimization modeling.

3.1 General Problem with All Indicators

Because his indicator declines with age, Hamilton deduced that mortality must increase with age. The relationship between his indicator of selection pressure and the age-pattern of mortality is not a simple one, however. During development his indicator is constant, while mortality, for many and perhaps all species, is falling. At post-reproductive ages his indicator is zero, while mortality, at least in humans, rises and then slowly levels off. Although the mismatch between indicator and pattern was acknowledged by Hamilton himself, an inverse relation between his indicator and the age-pattern of mortality is commonly assumed. The main justification, from Hamilton onwards, appears to be that there is an inverse relation between his indicator and the age-trajectory of mortality at reproductive ages in humans.

It is well known among plant biologists that many plants are capable of reducing their hazard of death by continued growth after the onset of reproduction. As discussed later in this chapter, various animals show negligibly increasing or declining mortality. I will show in Chaps. 4 and 5 that optimization models can lead to strategies where mortality is constant or keeps on falling after reproductive maturity. Figure 3.1 compares these patterns to Hamilton's inevitably decreasing indicator. It is clear that mortality is not necessarily an inverse function of Hamilton's indicator.

The alternative indicator that I suggested for the force of selection can increase with age, but only if the hazard of death is increasing. The indicator, however, can also decrease when the hazard of death is increasing: whether the indicator increases or decreases depends on how fertility is changing. Furthermore, the indicator decreases if the hazard of death is decreasing. So, as with Hamilton's indicator, the alternative indicator is not necessarily inversely related to the age-pattern of mortality.

But then how are the indicators of the force of selection against senescence related to the shape of the age-pattern of mortality? Hamilton quantified the selection pressure but he did not think carefully about the response to that pressure, although he acknowledges that "what way life schedules will be moulded by natural selection depends on what sort of genetical variation is available" [76, p. 118]. Lande [105] emphasizes that the change in a phenotype is determined by selection pressure (i.e. the indicator) together with the response matrix (the so called G-matrix), which includes variances and covariances for all fitness-relevant traits. The matrix not only takes into account "genetical variation" but also trade-offs among traits. Hamilton ignored these trade-offs.

The indicator of selection pressure together with the response matrix yields information about *short* term evolutionary processes. The implications for the long term, however, cannot be readily assessed because

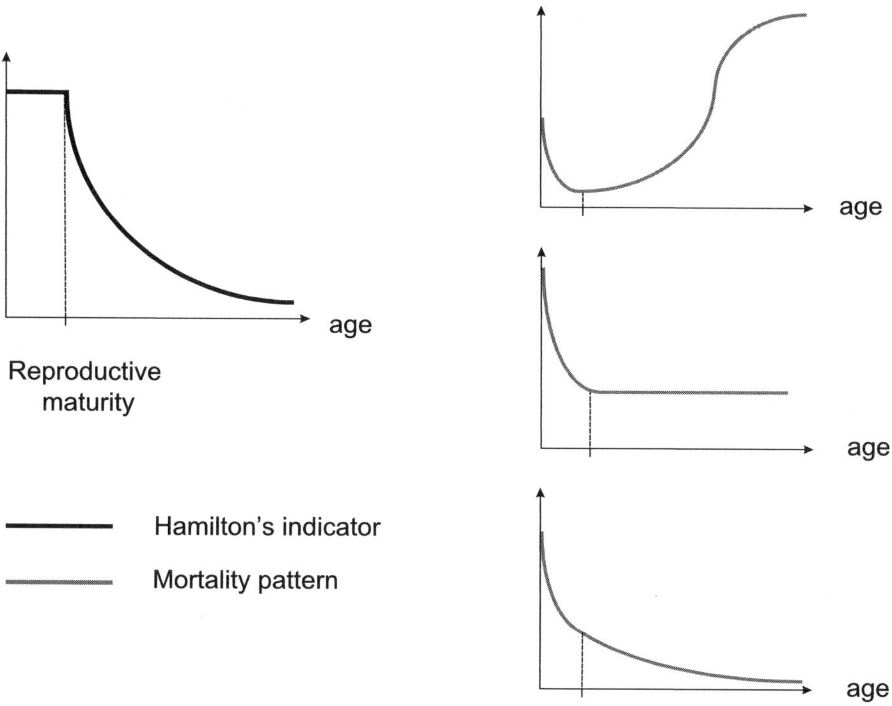

Fig. 3.1. The relation between Hamilton's declining indicator of selection pressure (*left side, in black*) and three possible age-patterns of mortality (*right side, in grey*)

the selection pressure is determined by what it shapes. The calculation of the indicator of selection pressure is based on the age-trajectories of mortality and fertility, and these trajectories depend on current levels of fitness-relevant traits. The entries in the G-matrix correspond to the variances and covariances at current levels of traits. But if, say, n traits are involved, then the indicator as well as the matrix take different values in an n-dimensional space. Evolution moves a species in this space at the speed and in the direction specifically determined by its position in that space. As position changes, speed and direction change.

In other words, as traits are shaped by evolution, they re-shape the selection pressure and possibly the G-matrix. It is not clear whether this process will ultimately converge and, if it does, to what evolutionary equilibrium. Since the force of selection is essential for evolutionary demographic theory, the implications of this feedback loop have to be understood. This requires modeling.

In sum, the quantities in Table 2.1 are *indicators* of the force of selection. They can provide an impression of the short-term direction and magnitude of the force of selection on age-specific survival and reproduction. But they are only one aspect of a multi-faceted story.

3.2 Theoretical Arguments

3.2.1 Mutation–Selection Balance

If mutation accumulation were the main explanation for senescence, which Hamilton assumes is a trait that is common to all individuals in a population, then each individual must be affected. For any particular deleterious mutation, mutation–selection balance implies that at least some individuals do not carry that mutation, namely those individuals in the zero-box. As long as selection pressure significantly exceeds mutation pressure, most individuals will be in the zero-box. Therefore, each individual would have to have his or her own set of deleterious mutations, being non-mutant for some genes and mutant for others. If genes had large and/or epistatic (non-linear) effects, a small set of genes could be sufficient. A population however, would then be highly heterogeneous, with some individuals suffering a rapid increase in mortality and others enjoying slow or postponed senescence. This does not appear to be the case, at least not for humans. Low variance in the age of senescent death requires the existence of many genes that have negative effects towards the end of reproductive life but no effects before that. Hamilton's theory assumes then that many genes have small effects that act additively. I will review the empirical evidence for age-specific, late-acting mutations in a subsequent section.

If there are few genes that have age-specific effects, then for mutation accumulation to be the main cause of senescence, these genes must be fixed in the population to lead to the phenomenon of senescence, which Hamilton claims to be universal. Fixation of a mutation implies that every individual in the population carries the same mutant allele for the gene in question. In this theoretical model this means that no individual is left in the zero box. The fixation of deleterious mutations at advanced ages poses a further challenge: unraveling.

3.2.2 Unraveling

Human mortality rises much more slowly than suggested by the results in Fig. 2.5, consistent with an earlier, similar observation by Abrams

[2, p. 357f]. This leads to a problem we have not yet touched on. All the indicators in Table 2.1 imply that the force of selection drops to zero when reproduction ceases. Several authors have argued that recurrent, deleterious mutations that only effect post-reproductive ages would become fixed, yielding a black hole of death at the age when reproduction ends [32, 146, 198, 208]. This could have been shown in all the figures above if the curves were drawn to higher ages. As $h_a^\dagger$ approaches ν, the equilibrium number of mutations steeply rises.

However, remaining reproduction $h_a^\dagger$ is calculated on the basis of a non-mutant life-history schedule. As $h_a^\dagger$ approaches zero, the equilibrium number of mutations rises to its maximum number at the age when $0 < h_a^\dagger << \nu$, even though a small fraction of reproduction is left. Hence, all bad genes after that age are fixed in the population and no individual is left with the non-mutant schedule. The disadvantage of carrying the mutation disappears, since every individual carries it. The fitness differential with respect to that mutation is gone. Therefore, a new $h_a^\dagger$ that falls more quickly near the end of reproductive life determines the selection pressure. Consequently, the point at which all mutations become fixed moves forward to a younger age. This process of unraveling would move the wall of death to younger and younger ages until it ultimately reaches maturity. Semelparity would be the only life-history strategy possible, which clearly is not the case.

Unraveling crucially depends on the age-trajectory of the mutation pressure. The age–window at the end of reproductive life, when selection pressure is weak and mutation pressure is strong, might be small. Let a be the first age when remaining reproduction is much smaller in magnitude compared to the mutation pressure, i.e. $0 < h_a^\dagger << \nu$, and A the age from which onwards remaining reproduction is zero, i.e. $h_A^\dagger = 0$. Unraveling will occur only if there are mutations whose effects become apparent inside but not before the age interval $[a, A]$. On the other hand, mutation accumulation will shape the age-pattern of mortality only if there are mutations that increase mortality at older ages, including within the interval $[a, A]$. Furthermore, such mutations cannot have major effects at ages where selection pressure is high.

In sum, there are many restrictions on the nature of the mutations that could permit mutation accumulation to shape aging, as discussed in this section and in the previous section. We will see in Sect. 3.3.1 below, that it is not clear that enough such mutations exist.

3.2.3 Variable Environments

Hamilton assumes a constant environment but environments are – in fact – variable. Accounting for changing environments can weaken mutation accumulation considerably. As the environment switches between good and bad times, it becomes essential during bad periods (during droughts, for example) to survive a long time in order to reproduce at all. Such a period would create a bottleneck. Only those individuals that were able to switch to a "survival mode", having no or very few bad mutations at higher ages, would constitute the gene-pool for all following generations, cleaning out any mutation accumulation.

Many species are able to switch between different life-history strategies depending on environmental conditions [7, 20, 63]. The same genome allows for strategies that can substantially differ in life expectancy. Given the short-lived strategy that might be optimal under average environmental conditions, mutations are predicted to accumulate at ages beyond the corresponding expected end of life. These mutations would raise mortality, preventing a substantial extension of lifespan, i.e. switching to the long-lived strategy. For species with alternative short and long-lived strategies, an increase in mortality with age in the short-lived strategy cannot be explained by mutation accumulation.

This reasoning only holds if mutations are assumed to be age-specific, i.e. time counting. Probably, however, gene expression is state-specific rather than age-specific. In this case a deleterious mutation could hide in the genome if the respective gene is not expressed in survival mode.

State– or condition-specific mutations could also explain results from an experiment conducted in Linda Partridge's laboratory. Mair et al. [116] show that dietary shifts can lead to switching between two different trajectories of mortality, one for the line on a restricted diet and one for the unrestricted line. The possibility of immediate shifts between a higher and a lower mortality curve in both directions, up and down, cannot be explained by simple mutation accumulation, especially since the shifts can occur at both younger and older ages. Such shifts and other kinds of plasticity in the age-pattern of mortality can, however, be explained by optimization models, as I discuss in Chap. 6.

Let me also note that the influence of unpredictable, stochastic environments (and in this regard also finite population sizes, finite time, and neutral theory) cannot be neglected when explaining the evolution of senescence [143, 199]. I will return to these points in Chap. 6.

3.2.4 Other Mechanisms

Variable environments are one counter–mechanism against mutation accumulation. Other mechanisms that can reduce the amount of mutations accumulating are synergistic epistasis and the occurrence of beneficial mutations [177, 211]. In the former case, the force of selection prevents the accumulation of mutations more strongly, because mutational effects magnify each other. In the latter case, the beneficial effect of some mutations offsets the deleterious effect of other mutations and therefore prevents an increase in mortality. Note that optimization of age-patterns of mortality, fertility and other traits results from the selection of beneficial mutations.

Hamilton pointed out that his results cannot explain the decline in mortality during development nor the existence of a post-reproductive period. Hamilton hypothesized that parental care is a missing piece in his framework that could account for both decreasing juvenile mortality as well as life after the end of reproduction. Parental care is a special form of resource transfer from parents to offspring. Lee, [109], Chu and Lee [36] and Robson and Kaplan [168] argue that intergenerational transfers that are made before, at and after birth can significantly influence the evolution of life-history schedules and, in particular, could explain the U-shaped trajectory of mortality in humans.

3.3 Empirical Evidence

3.3.1 Testing Preconditions for Mutation Accumulation

Three important preconditions for Hamilton's approach are:

- The existence of genes with effects confined to particular ages, especially to later ages.
- Mutations in these genes have small, deleterious effects.
- Effects of mutations do not interact with each other.

These preconditions have been tested empirically with an emphasis on the first condition.

To test the first precondition for the theory of mutation accumulation two large demographic studies in *Drosophila* have been conducted. Pletcher et al. [155] used inbred lines and found only weak evidence for the existence of mutations with deleterious effects confined to higher ages. The mutational load at later ages of their lines, however, might have been effectively saturated because of inbreeding depression (Yampolsky et al. [216], see also Sgrò and Partridge [178]). Negative epistasis at such a high mutational load could explain the results of Pletcher and colleagues [155]. Yampolsky and colleagues [216] conducted experiments with outbred lines of *Drosophila* and found clear evidence for age-specific effects after 10 and 20 generations. This evidence, however, decreased after 30 generations.

Evidence from Pletcher et al. [156] (for *Drosophila*) and Golden and Melov [67] (for *C. elegans*), who tested age-specific gene-expression levels, supports the existence of genes with age-specific effects, whereas Landis et al. [106] found a small tendency towards down-regulation of energy metabolism genes in *Drosophila* over adult ages. As a general pattern for both *Drosophila* and *C. elegans*, McCarroll et al. [124] found gene expression levels to be higher at younger ages than at later ages.

The second precondition of mutation accumulation is that mutations have small effects. Some mutations may, however, have major effects. It has been shown that the lifespan can be strongly effected by single mutations in *C.elegans* [89, 113] and *Drosophila* [40, 112, 145, 190].

Hamilton's third precondition is that aging-related genes should effect mortality in a linear, i.e. non-epistatic, manner. It has been shown, however, that genes effecting the lifespan of flies and worms interact [110, 180] and their expression depends on their genetic background [185]. Recently, Spencer and Promislow [184] showed for *Drosophila* that gene $\times$ genetic background interactions not only affect lifespan as a whole, but they also affect mortality in an age-specific manner. They conclude that aging-related traits could, to a significant extent, be shaped by age-specific epistasis. This possibility has not been considered so far in the evolutionary theories of senescence. The epistatic action of aging-related genes is further supported by Promislow [160], who shows that proteins associated with senescence interact more strongly than would be expected by chance.

If mutation accumulation were the main cause of senescence, the empirical evidence should be abundant and clear. The evidence, however, suggests that two out of three preconditions may be violated and evidence for the first precondition is not unambiguous.

3.3.2 Checking Predictions from Mutation Accumulation

If mutation accumulation were at work, then a main prediction is that there will be an increase in genetic variation and inbreeding effects with age. The evidence for an increase in genetic variation is mixed. Some evidence supports such an increase [82, 83] whereas others report an increase in genetic variance early in life followed by a decline in later life

[161, 191]. The strongest support for the mutation accumulation theory is given by Hughes et al. [84], who show a marked increase in both genetic variation and inbreeding effects in *Drosophila* with age. The authors emphasize that the increase in inbreeding effects is expected only under mutation accumulation, not under antagonistic pleiotrophy [30]. Caution should be exercised regarding evidence of increasing inbreeding depression with age because old flies may just be more enfeebled and hence susceptible to the effects of inbreeding.

On the basis of his results, Hamilton made predictions about the age-pattern of mortality. He inferred that mortality should be lowest at reproductive maturity and "trails upward indefinitely at the right ... roughly asymptotic to the age of the ending of reproduction" [76, p. 119], i.e. the theory of mutation accumulation would rule out the existence of a post-reproductive period. Mortality trajectories at older ages, however, have been found to level off and, in some studies, to decline for humans and various species kept in protected environments [21, 32, 47, 151, 201]. Several species studied in the laboratory have been shown to enjoy an extended period of post-reproductive life.

The level of extrinsic mortality determines the age beyond which remaining reproduction ($h_a^\dagger$) becomes negligible in the wild. This is the age at which Hamilton predicts a steep increase in mortality. The higher the extrinsic risk of death, the earlier the age at which mutations could accumulate. Hence, animals kept in laboratories, zoos, or other protected environments should suffer senescence at ages few of them would reach in the wild. Their lifespans should not exceed maximum lifespan in the wild. Many lab and zoo animals, however, live much longer than in the wild [19, 21]).

Furthermore, when kept protected from extrinsic hazards, a steeper rise in mortality with age is predicted for populations from high risk environments than for populations from lower risk environments. However, guppies from high risk pools showed a slower pace of senescence than guppies from lower risk pools when brought into the laboratory [163], contrary to the prediction of mutation accumulation theory. Differences in phenotypic development under high and low density conditions is one explanation for this phenomenon. Abrams [4] discusses this and several other explanations for the guppy puzzle. To explain long lives in protected environments, alternatives to the theory of mutation accumulation, e.g., alternatives based on optimization approaches, have to be found.

3.3.3 Empirical Evidence for Non-senescence

According to Hamilton senescence should be a ubiquitous characteristic of life histories, and mortality should start rising when reproductive maturity is reached. Three well-known gerontologists [43, 56, 188] emphasized, however, that "certain animals and plants do not manifest increases of mortality rate or other signs of senescence" [56, p. 221]. In particular, Finch [56, 57], Finch and Austad [58] and Ottinger et al. [144] have prepared the way for studies of non-senescence by focusing research on species with " negligible senescence", i.e., species for which death rates rise very slowly, if at all, with age. Caswell [23, p. 39] discusses increases in fertility as well as decreases in mortality with size (and therefore with age) and provides numerous examples and references.

The strongest evidence for non-senescence in animal species comes from studies of corals. Babcook [10] shows in three coral species (*Goniastrea aspera, G. favulus, and Platygyra sinensis*) that mortality is inversely related to colony size and age. Furthermore, the total fecundity of the three species increases steeply with size and age, "due to a combination of increased polyp fecundity and increased surface area"[10]. Grigg [70] presents comparable results for two other corals, *Muricea californica* and *Muricea fruticosa.*

Like the massive reef-building corals, some plants develop into large clonal clusters [56, Table 4.2, p. 229]. The quaking aspen (*Populus tremuloides*) grove studied by Kemperman and Barnes [93] covered 81,000 square meters and was estimated to be at least 10,000 years old. It seems likely that the bigger such a clonal cluster is, the lower is its chance of death.

Other species that are candidates for non-senescence include the wild leek *Allium tricocum* [136], brown algae *Ascophyllum nodosum* [1], the forest tree *Garcinia lucida* [71], the neotropical tree *Cecropia obtusifolia* [6] and the cushion plant *Limonium delicatulum* [78].

Strong evidence for a period of parallel increase in age-specific survival and fertility in non-modular animals can be found for some species of molluscs. Fertility often increases by ten-fold or so as individuals grow following reproductive maturity, and mortality decreases sharply (e.g., for the marine gastropods *Umbonium costatum* [139, 140] and *Littorina rudis* [85] and the bivalve *Yoldia notabilis* [134, 135]). There is also evidence of non-senescence for echinoderms such as sea urchins [53]. Hydra species [123] are likely candidates as well.

Some vertebrates may possibly enjoy non-senescence. Finch [56] summarizes suggestive data on rockfish, hagfish and various other species. For some reptiles, death rates decline somewhat after the age of reproductive maturity is reached, e.g., for *Sceloporus graciosus* [195], some populations of *Sceloporus undulatus* [194] and some populations of *Lacerta vivipara* [80][1].

Kohler et al. [100] analyze data sets for various species living in zoos and aquaria worldwide. They state that "there are several groups for which the age-pattern of mortality is nearly level". Comparing survival probabilities from the first decade of life (age 1 to 10, i.e. excluding juvenile death) with the second decade of life the evidence shows that raptors and crocodiles enjoy better survival in the second decade of their lives than in the first decade. Ratites show no signs of decrease in survival probability from their first to their second decade of life.

Non-senescent life histories cannot be explained by mutation accumulation.

3.4 Conclusion

The empirical evidence together with the theoretical arguments presented in this chapter indicate that mutation accumulation theory does not provide the fundamental explanation for the evolution of age-patterns of mortality. Together with my results from Chap. 2 they cast doubt on the assertion that senescence is inevitable.

It seems likely that the variety of possible age-trajectories of mortality is broad. Figure 3.2 summarizes various possibilities. During the first phase of life, development, mortality declines. During the second phase, mortality may increase, it may remain roughly constant, or it may decline. Then late in life, when most adults are dead, mortality may increase, level off or decline.

The age that marks the start and end of the different phases might be influenced strongly by growth patterns. For some species, growth ceases at reproductive maturity and marks the age when mortality starts rising. As noted above, however, individuals from many species continue to grow after the onset of reproduction and mortality may continue to fall until the age when growth stops. Models are needed to study which of these hypothetical age-patterns are theoretically possible. I will derive such models in the following two chapters.

---

[1] I thank my colleague Martin Dölling for his substantial help in gathering the references regarding evidence of non-senescent species.

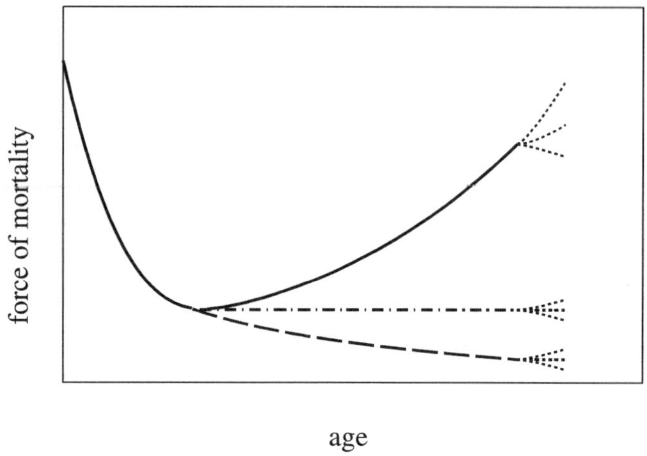

Fig. 3.2. Different hypothetical mortality trajectories

Note that the age-pattern of mortality reflects the average mortality in the population. The frail tend to die first. Hence, as individuals die, average mortality successively approaches the individual mortality trajectory of the most robust ones. The more heterogeneous a population is, the stronger is this effect. Therefore the age-pattern of mortality might exhibit a leveling and even a decline in mortality although the underlying individual age-pattern is still increasing [32, 151, 204, 206, 207].

The evidence suggests that mortality and fertility over the bulk of reproductive life are shaped by mechanisms other than mutation accumulation. Theories based on trade-offs might explain the existence of non-senescent life-history strategies [147, 148, 150]. It is not clear whether mutation accumulation plays a significant role in the evolution of senescence. If it turns out that mutation accumulation is an important mechanism for some species at older ages, then models of mutation accumulation need to be combined with trade-off models of the evolution of senescence to clarify the dynamics of demographic schedules [2, 200]. In the following two chapters, I develop trade-off models and explore their implications for the evolution of the age-patterns of mortality.

Part II

Optimization Models

4

Optimization Models Based on Size

4.1 Size Matters

Hamilton did not prove that senescence is inevitable. Furthermore, it seems likely that the age-trajectory of mortality is largely shaped by optimization: only at advanced ages, when the bulk of total lifetime reproduction has been realized, might mutation accumulation play a role. So the question arises: could it be optimal for a species not to follow a senescent life-history strategy?

As Caswell argues, for many organisms "the age of an individual tells little or nothing about its demographic properties" [23, p. 39]. Often what is important is size or stage of development. He concludes that " [s]ize-dependent demography is probably the rule rather than the exception and is especially pronounced in species with a large range of adult body size as a result of indeterminate adult growth."

Trees, for example, continue growing over an extended period of their life, gaining strength, becoming more robust and thereby reducing their susceptibility to death. (If trees at sites exposed to wind are too tall, then their susceptibility to damage and death might increase: this, however, is a special case.) A larger size (tall, thick stem, more leaves, longer roots) lowers the risk of death and enables better access to resources (light, water, nutrients). Larger trees produce more seeds than smaller trees.

The same is true for some species of fish. For instance, in some species the young adult fish, still small, has only a few progeny and is the prey of bigger fish. Over time the fish grows large enough to become a predator itself, increasing its level of resources and lowering its own risk of death.

Small alligators are prey to a variety of predators including raccoons, otters, wading birds, and fish. But most dangerous to small alligators probably are predators of their own kind, the larger alligators. Large alligators also die of cannibalism and fight with each other (see http://myfwc.com/gators/facts.htm). An individual alligator's size and strength determines whether it receives or becomes an additional ration of food.

In this chapter, I hypothesize that candidate species for non-senescent life histories are species that continue to grow substantially after the onset of reproduction and for which size is strongly associated with continued survival and reproductive success. This appears to be the case for the plant *Plantago lanceolata* after seasonal effects are removed [167]. The study of *Plantago lanceolata* by Deborah Roach was the particular motivation for me to develop a general life-history model based on size rather than age to understand whether non-senescence is theoretically possible.

Evidence for size-dependent mortality is reported for herbaceous plants in general [44, 77, 173, 182], thistles in particular [162, 171], trees [88], corals [86] and fish [129, 154]. Sauer and Slade [174] also document the effect of body mass on reproduction and survival in vertebrates.

For some species mortality may not decrease as size increases: there may be no relation, or mortality may increase with size. In addition, it is important to note that, for some species, larger size may not *cause* lower mortality. Larger size may have co-evolved with lower mortality, both resulting from some other aspect of the species' life history. For some species, for instance species like *Drosophila*, which exhibits discrete developmental stages rather than continuous growth, size may not be a key determinant of mortality. So a size-based model can shed light on the life history of only some species. But these species are "conceivable organisms" and may show non-senescent life-history strategies.

Size is the central state variable in the models I will develop in this chapter. Size determines mortality and fertility. Age enters the models only insofar as it takes time to grow – age itself does not matter. Using size as the state variable in these kinds of models is a first step to understanding whether *any* life history could be non-senescent. Note that the state variable size can be understood not only as physiological size but more generally as "size and strength". In Chap. 5, I develop a new model that is based on "vitality" rather than size.

4.2 A Size-Based Life-History Model

An optimal life history maximizes lifetime reproductive success. Accordingly, the energy available to an organism, which is always limited, has to be distributed among the basic processes of life: reproduction, maintenance and growth. How evolution solves this allocation problem determines the optimal trajectory of growth and thereby the optimal trajectories of the main demographic schedules, mortality and fertility.

All forms of life have to deal with damage. Damage occurs all the time and is discarded or repaired continuously, sometimes fully, sometimes partially. Models that take into account the influence of damage on mortality and fertility can do so on the occurrence and/or on the disposal and repair side. Energy allocation problems imply that disposal and repair of damage decreases when more energy is allocated to reproduction and therefore less energy remains for processes of maintenance and growth. Models based on the concept of energy allocation do not necessarily account for where the damage comes from. Reproduction itself, for instance, can be a direct cause of damage. For simplicity, the model I am going to develop in this chapter will focus on the energy allocation trade-off between reproduction, on the one hand, and maintenance and growth on the other. That is, I treat growth and repair as elements of the same general process and I do not explicitly model damage resulting from reproductive activities. I assume that the occurrence of damage increases proportionally with size.

Models based on the concept of optimal energy allocation over the life cycle represent a fundamental approach in life history modeling. Early applications of this concept were developed more than three decades ago, for example by Cole [42], Gadgil and Bossert [62], Schaffer [175], Taylor et al. [192], and Leon [111]. More recent examples of the application of the concept of optimal energy allocation include Charlesworth [26], Perrin [152], Perrin and Sibly [153], Kozlowski [102], Chichon [37], Teriokhin [193], Charnov et al. [35], Mangel and Stamps[122], Kaplan and Robson [91], Chu and Lee [36] and Charlesworth [34].

Generally, such life-history models are driven by the trade-off between reproduction and growth. Depending on the particular research focus, growth is sometimes further differentiated into growth of acquisition structure, storage structure, defense structure, reproductive structure and/or cognitive functioning. The central quantity of interest is the fraction of energy allocated to reproduction, the reproductive effort of an individual.

Life history models based on the concept of reproductive effort have been studied intensively (for a review see Charlesworth [27, Section 5.3.4.]). Common to these models is the assumption of a direct, inverse relation between survival and reproduction, which is mediated by reproductive effort. One outcome of these models is that reproductive effort should increase with age [62, 175]. However, Fragen [61] produced some counter-examples and Charlesworth and Leon [31] derived conditions that would lead to a decreasing reproductive effort with age, i.e. to an increase in survival with age. These results illuminate the general pattern of how reproductive effort should change with age. But, as Charlesworth [27, p 214] put it: "The problem of solving for the optimal life history with this model is a formidable one."

My research aim is to study the variety of qualitative patterns of mortality and fertility over age. In particular, I wish to understand whether it can be optimal for mortality to be constant or to fall over an extended period of life after the onset of reproduction. Interestingly, optimal patterns of mortality and fertility were commonly found to be flat in numerical studies by Charlesworth [26]. In these studies, reproductive effort increased so slowly, that it appeared to be virtually constant.

The examples given in the previous section suggest that, for some species, mortality decreases with size and fertility increases with size. For species with continued growth that follow this pattern, constant or falling mortality after the onset of reproduction seems to be optimal, at least for some period of the lifespan. Consequently, the models developed in this chapter are designed to capture this simple pattern based on the state variable size.

In contrast to previous reproductive effort models, the link between survival and reproduction will be mediated by size. The important implication of this assumption is that an increase in reproductive effort does not necessarily lead to a decrease in survival, and a decrease in reproductive effort does not necessarily lead to an increase in survival. I will emphasize this point in Sect. 4.2.2.

Every organism has to cope with the ubiquitous processes of deterioration. This means that some of the energy invested in " growth" is needed to repair damage. Only what is left after the requirements of maintenance have been met can be used to increase current size. Size changes according to the balance between repair and damage. Thus, size in this framework can increase, decrease or remain constant and, consequently, mortality can increase, decrease or remain constant. Whether mortality increases or decreases is an outcome of the model and not an assumption. This is a crucial feature, which distinguishes this model from previous models.

The importance of size is generally recognized [23, p.39]. A state-based model that assumes an inverse relation between state and mortality has been developed before by Perrin [152]. However, Perrin implicitly assumes a non-senescent life history because mortality cannot increase in his model. Perrin's approach does not account for the occurrence of damage and its possible repair. A model that incorporates damage and repair was developed by Kozlowski [102], Cichon [37] and Cichon and Kozlowski [39]. In their framework mortality does not depend on state but on accumulated damage and can, at best, remain constant. Complete repair of current damage is realized only if all energy is invested in repair, i.e. at the cost of zero reproduction. Otherwise mortality rises at a pace determined by reproductive effort. An increase in mortality is inevitable.

An innovative feature of the approach I will be taking is that I combine the inverse relation of mortality and size with the possible accumulation of damage and its repair. My research builds on and further develops Vaupel, Baudisch et al. [200]. Mangel and colleagues [120, 121] have recently developed other models in which mortality is the consequence of growth and metabolism and associated damage.

4.2.1 The General Optimization Problem

The general optimization problem can be formalized as follows. Let $\xi(a)$ denote the size (and strength) of an individual at age a. Let $\pi(a)$ denote the fraction of energy allocated to growth at that age. Assume that the change in size over age depends on investment $\pi(a)$ and size $\xi(a)$ but not on age a itself, i.e. that the trajectory of $\xi(a)$ is determined by the autonomous first-order differential equation

$$\frac{d\xi}{da} \equiv \dot{\xi} = g(\xi(a), \pi(a)) \ . \tag{4.1}$$

Note that the dot indicates a change over age. Initial size is given by $\xi(0)$. From that size onwards, the age-trajectory of $\pi(a)$ determines the age-trajectory of $\xi(a)$.

The optimal trajectory of $\pi(a)$ over the life course is assumed to be the strategy that maximizes Darwinian fitness, measured as lifetime reproductive success, a functional of the form

$$\max R = \int_0^\infty f(\xi(a), \pi(a)) \, da \ , \tag{4.2}$$

where $f(\xi(a), \pi(a))$ depends on the age-trajectories of mortality and fertility and hence on the age trajectories of $\xi(a)$ and $\pi(a)$. The age horizon is potentially infinite, but non-zero mortality insures that every individual has a finite lifespan.

The general optimization problem is described by the objective as given in (4.2) and the autonomous first-order differential equation as given in (4.1), which determines the change in size over age.

4.2.2 The Specific Optimization Problem

The change in size is determined by the fraction of energy invested in growth, $\pi(a)$. Energy is allocated between growth and maintenance on the one hand, and reproduction on the other hand. The fraction of energy allocated to reproduction, the reproductive effort, is captured by $1 - \pi(a)$, since in this model maintenance and growth are assumed to be paid out of the same budget. In accordance with the literature, the change in size is assumed to be inversely related to reproductive effort.

Larger size implies higher complexity, which is more costly to maintain. The rate of occurrence of new damage will be assumed to increase proportionally with size [101, 210]. A simple way of modeling deterioration is to assume a linear relation with size, i.e.

$$\delta(\xi(a)) = \delta_0 + \delta_1 \xi(a) , \tag{4.3}$$

where $\delta_0 > 0$ and $\delta_1 > 0$ are constant parameters.

Size is assumed to change proportionally to the level of current size $\xi(a)$. This implies the assumption that available resources are proportional to size, an assumption also made by Charlesworth and Leon [31], Gadgil and Bossert [62] and Leon [111]. Furthermore I assume that the change in size is proportional to the difference between investment $\pi(a)$ and deterioration $\delta(\xi(a))$. Growth only occurs if investment exceeds the current rate of deterioration. Therefore, the change in size can be specified as

$$\frac{d}{da}\xi(a) = k\,(\pi(a) - \delta(\xi(a)))\,\xi(a) \tag{4.4}$$

where $k > 0$ is a constant scaling parameter. Initial size can be normalized by setting $\xi(0) = 1$. Substituting (4.3) into (4.4) yields the following logistic differential equation

$$\frac{d\xi(a)}{da} = k\,(\pi(a) - \delta_0 - \delta_1 \xi(a))\,\xi(a) . \tag{4.5}$$

This equation captures the change in size and specifies the general function $g(\cdot)$ of (4.1).

Life starts off with growth. Then at some age some energy is invested in reproduction. This age at onset of reproduction (reproductive maturity) α is determined by the age when $\pi(a) < 1$ for the first time. Figure 4.1 depicts the age-trajectory of size during development. The curve is given by the solution to (4.5), namely

$$\xi(a) = \left(\frac{\delta_1}{1 - \delta_0} + \left(1 - \frac{\delta_1}{1 - \delta_0}\right) e^{-k\,(1 - \delta_0\,)\,a} \right)^{-1} , \qquad (4.6)$$

taking into account that investment is constant at $\pi(a) = 1$ over that period and $\xi(0) = 1$. This logistic function has an upper limit of $(1 - \delta_0)\,/\,\delta_1$, which reflects the size an organism would eventually approach if it continues to spend all available resources on maintenance and growth. In size-based approaches, growth functions that have an

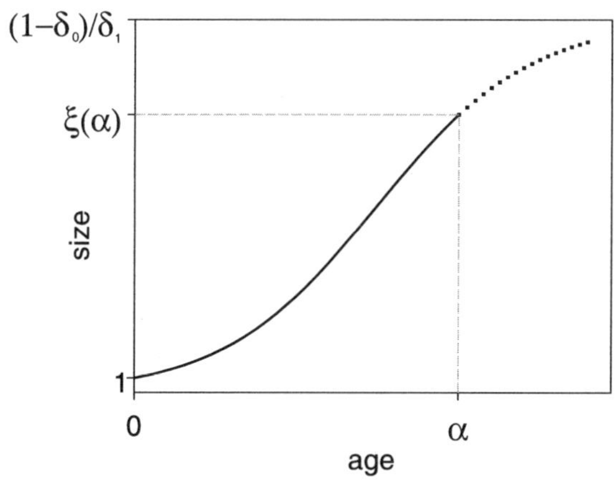

Fig. 4.1. Size $\xi(a)$ as a function of age a according to (4.5)

upper bound, such as the logistic function or the von Bertalanffy growth function, are frequently used, since size cannot increase indefinitely.

To ensure that the initial investment of $\pi_0 = 1$ actually leads to growth an additional restriction on the parameters in (4.3) is necessary. From (4.5) one gets

$$\left.\frac{d\xi(a)}{da}\right|_{a=0} = k\,(1 - \delta_0 - \delta_1) > 0$$

and hence

$$\delta_0 + \delta_1 < 1 \,. \tag{4.7}$$

This inequality concurrently guarantees that $\delta(\xi) < 1$.

The general function $f(\cdot)$ given in (4.2) can be specified by the product of the probability of surviving to age a, $l(a)$, and the amount of reproduction at that age, $m(a)$. The objective function is then specified by

$$\max R = \int_0^\infty l(a)\, m(\,\xi(a),\, \pi(a)\,)\; da \,. \tag{4.8}$$

The survival function $l(a)$ is determined by the trajectory of mortality up to age a via

$$l(a) = e^{-\int_0^a \mu(\xi(t))\, dt} . \tag{4.9}$$

The age-specific force of mortality, denoted by $\mu(a)$, is assumed to be inversely proportional to $\xi(a)$. As discussed in Sect. 4.1, I focus on species for which growth enhances future survival. A simple way to model mortality in this case is to let

$$\mu(a) = \frac{b}{\xi(a)} + c \,. \tag{4.10}$$

The constant parameter $b \geq 0$ captures the size-dependent, "intrinsic" component of death and the constant parameter $c > 0$ captures the size-independent, "extrinsic" component of death.

The model implies that, if no energy is allocated to growth, then size deteriorates exponentially and therefore mortality increases exponentially. However, whether it is optimal to invest all available energy in reproduction is an outcome of the model. An exponential increase in mortality is not a built-in property of the model. If mortality increases, it can do so at any pace, exponential being the extreme case. In the exponential case, the mortality function is the same as the Gompertz-Makeham function. Exponentially increasing mortality ("Gompertz Law") is frequently assumed in the literature, based on various empirical observations. The general structure of the mortality function is the same as that used by Perrin [152] (except for an exponent to size).

In accordance with the literature, I assume reproduction to be proportional to available resources (which are proportional to $\xi(a)$) and to the reproductive effort (in this model $(1-\pi(a))$). A simple way to specify reproduction is to assume a linear relation with reproductive effort; this approach was taken by Charlesworth [26], Perrin [152], Kozlowski

[102], Cichon [37] and Cichon and Kozlowski [39]. The maternity function, denoted by $m(a)$, is thus given by

$$m(a) = \varphi\,(1 - \pi(a))\,\xi(a)\,. \tag{4.11}$$

Note that the constant, positive parameter φ can be adjusted to ensure that the optimal strategy yields a net reproduction rate $R = 1$. This implies that population density is assumed to affect lifetime reproductive success in a proportional manner. Note further that fertility and mortality are written as functions of age for purposes of brevity only. To be precise, $m(a) = m(\xi(a), \pi(a))$ and $\mu(a) = \mu(\xi(a))$.

The pleiotropic effects of size can be summarized as

$$\frac{d}{d\xi}\,\mu(\xi) < 0, \qquad \frac{\partial}{\partial\xi}\,m(\xi,\pi) > 0, \qquad \frac{d}{d\xi}\,\delta(\xi) > 0\,. \tag{4.12}$$

A larger size implies a lower risk of death, a higher reproductive potential but also a higher level of deterioration, which increases the costs of maintenance. Recall that the mediating effect of size between mortality, fertility and damage constitutes an important difference to previous models of reproductive effort, as emphasized at the beginning of this chapter. Equations (4.3), (4.4), (4.10) and (4.11) imply that an increase in reproductive effort $(1-\pi(a))$ does not necessarily lead to a reduction in survival. As long as the level of $\pi(a)$ does not fall below the level of damage $\delta(\xi(a))$, size does not shrink and therefore mortality does not increase. Conversely, a declining investment in reproduction does not lead to improved survival as long as the level of investment $\pi(a)$ is below the level of damage $\delta(\xi(a))$.

4.3 An Optimization Model that Leads to Non-senescence

The optimal solution is a trajectory over age. Therefore, this is a dynamic rather than a static optimization problem. Two main approaches can be distinguished: Bellman's dynamic programming approach [12] and Pontryagin's Maximum Principle [157]. Comprehensive treatments of dynamic programming methods applied to biological problems are given in Mangel and Clark [118] and Clark and Mangel [41] as well as Bulmer [16]. The Appendix to Mangel [119] shows how to connect dynamic state variable modeling with the ideas of classical demography and life history models.

4.3.1 The State Ratchet

Bellman's general way of thinking implies a feedback loop strategy. In any particular given state, make the best possible decision. This decision will steer the state to some subsequent level. Again, given this subsequent state, do the best you can do. An optimal trajectory of decisions can be found by beginning at the last possible state and working backwards. The most important precondition for this strategy is that decisions only depend on the current state and potential future gains and losses but not on the past.

In particular, at each size $\xi(a)$ the amount of energy invested in growth $\pi(\xi(a))$ at that size determines whether size increases, decreases or is maintained. Depending on this decision, size changes over age according to (4.2). The optimal trajectory of energy allocation to growth determines the optimal trajectory of size over age, which in turn determines the optimal age-trajectories of mortality and fertility.

Following Bellman's way of reasoning, the general nature of the optimal strategy can be understood intuitively. Assume each size is associated with a unique level of optimal investment and size changes continuously over age. Then each $\xi(a)$ is associated with a single $\pi^*(a)$ (the star indicating 'optimal') that determines whether size increases, decreases or is maintained.

Assume at a particular size $\xi(a)$ that the optimal investment results in an increase in size to $\xi(a^+) > \xi(a)$ at age $a^+ > a$. Assume further that, at the subsequent bigger size, it would be optimal to shrink. Then size would shrink to some lower value $\xi(a^{++}) < \xi(a^+)$ at age $a^{++} > a^+$. However, size is a continuous variable. In order to grow from $\xi(a)$ to $\xi(a^+)$ it must have been optimal to grow at each intermediate size between $\xi(a)$ and $\xi(a^+)$. Shrinking again from $\xi(a^+)$ to $\xi(a^{++}))$ would imply that this optimality is violated at each level of size between $\xi(a^+)$ and $\xi(a^{++})$. Each intermediate size would be associated with two optimal strategies instead of one, which is a contradiction.

This line of reasoning leads to an important result, which I will call "the state ratchet". If, for the optimization problem formulated above, an optimal solution exists and each state is associated with exactly one optimal strategy, then any continuous, optimal state trajectory must be a monotonic function over age. Consequently, if the state variable initially increases, it will never decrease and if the state variable initially decreases it will never increase. Since maintenance implies that state does not change, the optimal strategy, which is bound to state only, will not change over age. Therefore, if, for any finite interval, it is optimal to maintain the current state, it will be maintained forever.

The state ratchet has important consequences for any optimal life history in this framework. Since life begins with growth it can never be optimal to shrink. Size can only increase and then be maintained at some point. Since mortality is assumed to be inversely related to size, mortality can never increase. Senescence is impossible. Intriguingly, this simple approach challenges Hamilton's postulate of inevitable senescence. It is possible to overcome the state ratchet, as I will discuss in a later section of this chapter, but only by making the model more complicated. Let's first consider the basic model.

4.3.2 The Maximum Principle

Pontryagin's way of thinking involves planning the whole future at time zero, in contrast to Bellman's backward step-by-step approach. Optimizing all future decisions at time zero requires knowledge about how decisions, the "control variable(s)", influence the change in the state variable(s) over time. The change in state(s) over time is determined by the so called "equation(s) of motion", i.e. first order differential equations that capture the change in any state variable over age. For my particular problem the control variable is the investment in growth, $\pi(a)$. One state variable is size, $\xi(a)$. Equation (4.5) determines the corresponding equation of motion, the change in size over age.

As in Bellman's approach, there is an important precondition. The optimal decision at any age a should only depend on the current state and potential future gains and losses but not on previous ages. However, survival to age a, as given in (4.9), depends on the trajectory of mortality between age zero and age a. Therefore, survival must be treated as an additional state variable. Note that survival changes over time according to

$$\frac{d}{da}\, l(a) \;=\; -l(a)\, \mu(a) \tag{4.13}$$

with initial condition $l(0) = 1$. Equation (4.13) depicts the equation of motion for the second state variable, survival.

Pontryagin's Maximum Principle [157] associates a specific function with the optimal control problem stated above, the "Hamiltonian"

$$\begin{aligned} H(\xi, l, \pi, \lambda_1, \lambda_2) \;=\; & l(a)\, m(\xi, \pi) \\ & + \lambda_1(a)\left[k\left(\pi(a) - \delta_0 - \delta_1\, \xi(a) \right) \xi(a) \right] \\ & - \lambda_2(a)\, l(a)\, \mu(\xi)\,. \end{aligned} \tag{4.14}$$

The first term is the contribution of the objective function (given in (4.8) at age a: This term captures the current gains from a decision

$\pi(a)$ at states $\xi(a)$ and $l(a)$. The remaining terms are the weighted sum of the change in the state variables. The factors $\lambda_1(a)$ and $\lambda_2(a)$ are costate variables. Costate variables capture the values of a hypothetical additional unit of $\xi(a)$ and $l(a)$ respectively at age a, the "shadow price" of size and survival.

Conditions for an Optimum

The Maximum Principle requires that an optimal solution necessarily fulfills the following criteria:

- The Hamiltonian function is maximized with respect to the investment strategy. In general, if $H(\cdot)$ is differentiable, then

$$\frac{d}{d\pi} H(\cdot) = 0 . \tag{4.15}$$

  In particular

$$H_\pi(\cdot) = l(a)\, m_\pi(\xi, \pi) + \lambda_1(a)\, k\, \xi(a) = 0 , \tag{4.16}$$

  the subscript π indicating the partial derivative. Clearly, if the Hamiltonian is linear in the control variable, then the maximum is attained at the boundaries of the feasible set for the control. Note that the last term dropped out. The shadow price of survival does not influence the maximum of the Hamiltonian.
- Furthermore the "adjoint equations"

$$\frac{d}{d\xi} H(\cdot) = -\frac{d}{da} \lambda_1(a), \quad \text{and} \quad \frac{d}{dl} H(\cdot) = -\frac{d}{da} \lambda_2(a) \tag{4.17}$$

  must hold. The change in the shadow price of a state variable must equal the negative change in the Hamiltonian with respect to that state. More specifically, the adjoint equations associated with size and survival, respectively, are given by

$$\begin{aligned} \dot{\lambda}_1 &= -H_\xi(\cdot) \\ &= -\, l\, m_\xi(\xi, \pi) \\ &\qquad - \lambda_1\, k\, (\pi - \delta_0 - 2\delta_1 \xi) \\ &\qquad\quad + \lambda_2\, l\, \mu_\xi(\xi) \end{aligned} \tag{4.18}$$

  and

$$\dot{\lambda}_2 = -H_l(\cdot) = -\, m(\xi, \pi) + \lambda_2\, \mu(\xi) . \tag{4.19}$$

- As age approaches infinity the values of an additional unit of size and survival, as captured by λ_1 and λ_2, respectively, have to approach zero. This is reflected in the transversality conditions, given by

$$\lim_{a \to \infty} \lambda_1(a) = \lim_{a \to \infty} \lambda_2(a) = 0 . \tag{4.20}$$

Note that the state, control and costate variables are all functions of age. However, for brevity they are written as ξ, π, λ_1 and λ_2 wherever no confusion arises.

Solution

Taking into account that

$$k(\pi - \delta_0 - 2\delta_1 \xi) = \frac{\dot{\xi}}{\xi} - k\delta_1 \xi \tag{4.21}$$

the solution to the differential Equation in (4.18) gives the shadow price of an additional unit size at age a,

$$\lambda_1(a) = -\frac{1}{\xi(a)} \int_a^\infty e^{-k\delta_1 \int_a^t \xi(\tau)\, d\tau}\, \xi(t) \times l(t)\, (\lambda_2(t)\, \mu_\xi(t) - m_\xi(t))\, dt . \tag{4.22}$$

Equation (4.19) can be solved as

$$\lambda_2(a) = \frac{1}{l(a)} \int_a^\infty l(t)\, m(t)\, dt . \tag{4.23}$$

The shadow price of survival at age a is equivalent to the reproductive value at that age. Inserting (4.23) into (4.18) leads to

$$\lambda_1(a) = \frac{1}{\xi(a)} \int_a^\infty e^{-k\delta_1 \int_a^t \xi(\tau)\, d\tau}\, \xi(t) \times \left(l(t)\, m_\xi(t) - \mu_\xi(t) \int_t^\infty l(\tau)\, m(\tau)\, d\tau \right) dt . \tag{4.24}$$

To find an explicit expression for size, (4.5) can be solved, resulting in

$$\xi(a) = \frac{\exp\left\{\int_0^a k\,(\pi(t) - \delta_0)\, dt\right\}}{\frac{1}{\xi(0)} + \int_0^a k\,\delta_1 \exp\left\{\int_0^t k\,(\pi(\tau) - \delta_0)\, d\tau\right\} dt} . \tag{4.25}$$

It can be seen that the state variable size increases in a logistic manner.

Result

With the state ratchet I showed that size must follow a monotonic path. The same result can be proved applying optimal control theory. For an infinite horizon autonomous optimal control problem with a single state variable, the optimal state path must be monotone (Kamien and Schwartz [90, p. 179] and Léonard and Van Long [114, p. 294]). Recall from (4.11) that fertility is linear in π. Therefore, the Hamiltonian function is linear in π, which results in solutions at the boundaries of the feasible set of investment strategies π, i.e. either one or zero.

Initially, $\pi_0 = 1$ and π remains at one until maturity. At maturity, a boundary solution implies that $\pi = 0$. If this were so, size would decrease, contradicting the state ratchet. Therefore, one expects what is called a "singular solution" in control theory. A singular solution requires that

$$\dot{H}_\pi = 0 = \dot{l}\, m^*_\pi + \dot{\lambda}_1\, k\, \xi^* \tag{4.26}$$

has to be satisfied. It would be natural if $\pi = \delta(a)$ were the singular solution required. Since size is constant in maintenance mode, the optimal solution would stay on the singular path forever. It turns out that $\pi = \delta(a)$ is the singular solution, as discussed below.

Since a logistic increase in size implies an upper limit to growth, there must be an age a^* at which size is finally maintained,

$$\pi = \delta(\xi), \qquad \forall\, a \geq a^* . \tag{4.27}$$

Consequently $\xi(a^*) = \xi^*$, $m(a^*) = m^*$ and $\mu(a^*) = \mu^*$ will be constant. If size is constant the reproductive value is simply given by the quotient of m^* and μ^*. Since the reproductive value of an individual at age a is captured by the costate variable $\lambda_2(a)$, this costate will be constant as well.

Assume $\pi = \delta(a)$ from age a^* onwards. Taking into account that

$$l(a) = l(a^*) e^{-\mu^*\,(a - a^*)}, \tag{4.28}$$

it follows from (4.24) for all $a \geq a^*$ that

$$\lambda_1(a) = \frac{l(a)\, m^*}{(k\, \delta_1\, \xi^* + \mu^*)} \left(\frac{m^*_\xi}{m^*} - \frac{\mu^*_\xi}{\mu^*} \right) . \tag{4.29}$$

This expression combined with condition (4.26) leads to an equation that determines the size at which the optimal investment should switch to maintenance mode,

$$\frac{m^*_\pi}{m^*} = \frac{k\,\xi^*}{k\,\delta_1\,\xi^* + \mu^*}\left(\frac{\mu^*_\xi}{\mu^*} - \frac{m^*_\xi}{m^*}\right). \tag{4.30}$$

The relative change in reproduction with respect to the investment in growth must equal the weighted difference between the relative changes in mortality and reproduction with respect to size. Note that this condition does not depend on age: (4.26) will be zero for all ages $a > a^*$ once maintenance mode is reached.

In this model fertility is given by (4.11). From (4.30) it follows that a singular solution is determined by

$$\frac{\mu(\xi^*_a)}{k} = (1 - \delta_0 - 2\,\delta_1\,\xi^*_a) + \frac{(1 - \delta_0 - \delta_1\,\xi^*_a)\,b}{\mu(\xi^*_a)\,\xi^*_a}. \tag{4.31}$$

The individual will grow at full speed until its size satisfies (4.31)[1].

Substituting $\mu(\xi) = b/\xi + c$ yields a cubic polynomial with three roots. Generally, these roots can be real and complex. Viable strategies correspond to real, nonnegative roots. The optimal size at maturity corresponds to the root that maximizes life-time reproduction. Strategies can be determined numerically; I used MATHEMATICA™ to calculate the solution.

4.3.3 An Alternative Derivation

The state ratchet implies that if there is a single state variable, then the optimal investment strategy of an organism has to be growth, possibly followed by maintenance, i.e. the feasible set of $\pi(a)$ is

$$\pi(a) \in [\,\delta(a),\, 1\,]\,. \tag{4.32}$$

A valuable hint follows from Pontryagin's Maximum Principle. Since the Hamiltonian is linear in $\pi(a)$ the optimal investment maximizes the Hamiltonian function at the boundaries of the feasible set (4.32). The upper limit $\pi(a) = 1$ is associated with full growth and no reproduction. The lower limit $\pi(a) = \delta(a)$ switches the organism to maintenance mode with constant, nonzero fertility and mortality.

[1] I thank Anatoli Michalski for his explanations regarding optimal control theory.

In this case the integral in (4.8) can be solved explicitly. The switching age, when $\pi(a)$ drops to $\delta(a)$, marks the onset of reproduction, age α. It follows that

$$R = l(\alpha)m(\alpha) \int_{\alpha}^{\infty} \exp\left\{-\int_{\alpha}^{a} \mu(t)\,dt\right\} da = l(\alpha)\,\frac{m(\alpha)}{\mu(\alpha)}\,, \qquad (4.33)$$

where $m(\alpha)$ and $\mu(\alpha)$ are the constant levels of fertility and mortality in maintenance mode after α.

The age α at which reproduction starts is determined by the value ξ_α that maximizes R in (4.33). Using the fact that from age zero to α there is a one-to-one correspondence between age a and size ξ, one can express (4.33) as a function of ξ_α. Inverting the logistic growth function $\xi = L(a)$ given in (4.6) leads to

$$a = L^{-1}(\xi) = \frac{1}{k\,(1 - \delta_0)}\,\ln\left(\frac{1 - \frac{\delta_1}{1-\delta_0}}{\frac{1}{\xi} - \frac{\delta_1}{1-\delta_0}}\right). \qquad (4.34)$$

Thus, by substituting $\alpha = L^{-1}(\xi_\alpha)$ in (4.33) one can express $R = R(\xi_\alpha)$ as a function of size at reproductive maturity ξ_α. The optimization problem now can be solved by setting the derivative of $R(\xi_\alpha)$ with respect to ξ_α equal to zero, i.e.,

$$l_{\xi_\alpha}\,\frac{m}{\mu} + m_{\xi_\alpha}\,\frac{l}{\mu} - \mu_{\xi_\alpha}\,\frac{l\,m}{\mu^2} = 0\,. \qquad (4.35)$$

Because

$$l_{\xi_\alpha} = \frac{d}{d\xi_\alpha}\,l(\xi_\alpha) = \frac{d}{d\xi_\alpha}\,\exp\left\{-\int_1^{\xi_\alpha} \mu(\xi)\,[k\,(1 - \delta_0 - \delta_1\,\xi)\,\xi]^{-1}\,d\xi\right\}$$

$$= -l(\xi_\alpha)\,\mu(\xi_\alpha)\,[k\,(1 - \delta_0 - \delta_1\,\xi_\alpha)\,\xi_\alpha]^{-1}\,,$$

optimal size at maturity is given by

$$\frac{\mu(\xi_\alpha)}{k} = (1 - \delta_0 - 2\,\delta_1\,\xi_\alpha) + \frac{(1 - \delta_0 - \delta_1\,\xi_\alpha)\,b}{\mu(\xi_\alpha)\,\xi_\alpha}\,. \qquad (4.36)$$

This equation is equivalent to (4.31). Using calculus and static optimization and applying Bellman's way of thinking with a hint from Pontryagin leads to the same result as using dynamic optimization applying Pontryagin's Maximum Principle.

4.3.4 The Simplest Model Leads to Sustenance

In the simplest case of size-independent mortality, i.e. $b = 0$, an explicit solution for the optimal size at maturity can be derived:

$$\xi_\alpha = \frac{(1 - \frac{c}{k} - \delta_0)}{2\,\delta_1} . \tag{4.37}$$

Results for three illustrative parameter combinations are shown in Fig. 4.2. Equation (4.37) implies

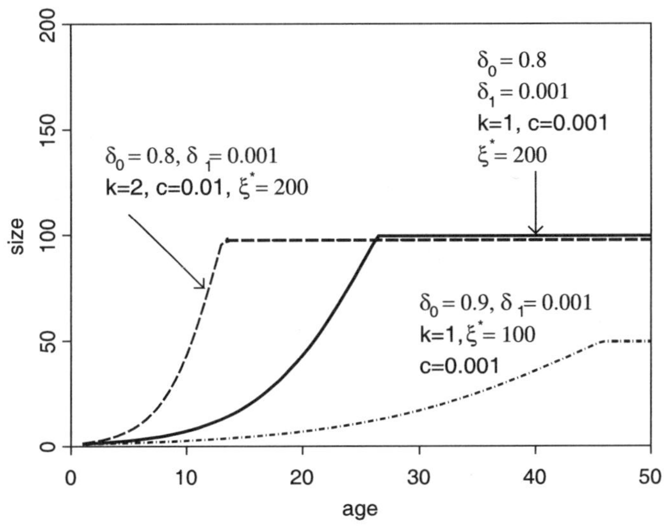

Fig. 4.2. Size $\xi(a)$ for three selected parameter combinations (Note that ξ^* denotes maximum possible size.)

$$\frac{d\xi_\alpha}{dc} < 0\,, \; \frac{d\xi_\alpha}{d\delta_0} < 0\,, \; \frac{d\xi_\alpha}{d\delta_1} < 0 \text{ and } \frac{d\xi_\alpha}{dk} > 0\,. \tag{4.38}$$

Furthermore, (4.37) and (4.34) imply

$$\frac{d\alpha}{dc} < 0 \text{ and } \frac{d\alpha}{d\delta_1} < 0\,. \tag{4.39}$$

Increasing extrinsic mortality reduces age and size at maturity. Changes in α with respect to k and δ_0 depend on the parameter combination

Table 4.1. Optimal size ξ_α and age α at the start of reproduction for size-dependent mortality ($b > 0$) according to (4.31)

ξ_α	α	ξ_{max}	$l(\alpha)$	b	c	k	δ_0	δ_1
62.26	50.96	100	0.005	0.5	0.001	1	0.9	0.001
53.46	47.34	100	$1.1 \cdot 10^{-9}$	2	0.001	1	0.9	0.001
60.02	50.02	100	0.00003	1	0.001	1	0.9	0.001
25.68	17.66	100	0.0012	1	0.1	2	0.9	0.001
56.86	24.36	100	0.0045	1	0.01	2	0.9	0.001
64.06	25.87	100	0.0056	1	0.000001	2	0.9	0.001
127.66	29.31	200	0.006	1	0.001	1	0.8	0.001
129.18	14.74	200	0.08	1	0.001	2	0.8	0.001

in a rather complicated way. For very small maximum attainable sizes and very slow speed of growth, α can increase with increasing k and decrease with increasing δ_0. Usually, however, an increase in k will lead to a decline in α while an increase in δ_0 will lead to a decrease in α.

If $b > 0$ in (4.10), then mortality declines as size increases. Hence for positive but small b

$$\xi_\alpha\,|_{b>0} > \xi_\alpha\,|_{b=0}\ . \tag{4.40}$$

If, however, b is large then the increased risk of death may make it optimal to start reproducing at a smaller size. Some illustrative results are shown in Table 4.1. If b gets too large then the resulting solutions are nonviable strategies: the species cannot survive because mortality is too high. Such nonviable strategies correspond to roots of (4.31) that are complex or negative.

In sum, the simplest model in which a single state variable determines the optimal strategy and reproductive effort affects fertility in a linear way can only lead to sustenance, i.e. a period of development followed by maintenance. Senescence is impossible and all there is to be optimized is the age at maturity. From this age onwards the individual maintains its state forever. Complications have to be added to the simple model to get optimal strategies that are more flexible than this basic strategy. Note that flat mortality and fertility profiles were found to be very common in numerical studies by Charlesworth [26].

4.3.5 Introducing Nonlinearity Can Lead to Enhancement

Enhancement[2] – a sustenance strategy that includes a period of parallel growth and reproduction after the initial period of development and before the terminal period of maintenance – is precluded by the linearity in $\pi(a)$ of Pontryagin's Hamiltonian. To allow enhancement a model specification has to be found which results in a Hamiltonian that is nonlinear in $\pi(a)$.

To solve such an optimization problem the Bellman principle of dynamic programming can be used. Because the size ratchet precludes an organism from returning to previous states, the optimal trajectory of the allocation strategy can be found by a backward algorithm starting at the maximum attainable size at which maintenance is the only possible strategy. I developed such an algorithm, which produced results that were consistent with the analytic solution in the case of fertility being linear in $\pi(a)$. This algorithm can be readily applied to the following nonlinear fertility function:

$$m(a) = \varphi\, \pi(a)\, (1 - \pi(a))\, \xi(a) = \varphi\, (\pi(a) - \pi^2(a))\, \xi(a)\,. \quad (4.41)$$

The second term in the product, $\pi(a)$, can be interpreted as the efficiency of converting size $\xi(a)$ into reproduction $m(a)$. As $\pi(a)$ approaches zero, i.e. as resources are largely directed to fertility rather than growth and maintenance, this efficiency declines.

Figure 4.3 shows an illustrative result. For the parameters used in this model, reproduction starts when the organism grows to about 25% of its potential maximum size. Then, until maintenance mode is eventually reached at age 250, there is an extended period of enhancement.

This still simple model leads to optimal strategies of development followed by a period of parallel growth and reproduction followed by

[2] Maren Rebke and James W. Vaupel suggested this term to describe a period of life with increasing fertility and declining mortality. This enhancement is due to some kind of growth, but perhaps in strength or capability and not in size. I also use it to describe a life history strategy that starts with development, switches to parallel growth and reproduction and then ends with maintenance.

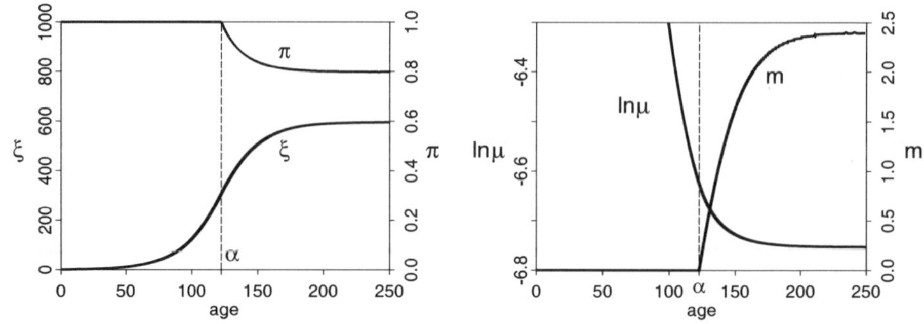

Fig. 4.3. Enhancement for model variant (4.41) (Parameter values were $k = 0.1$, $\delta_0 = 0.5$, $\delta_1 = 0.0005$, $b = 0.1$, $c = 0.001$, $\varphi = 0.02$. The force of mortality before age 100 is very high and rapidly falling.)

maintenance. In addition to the age at maturity, the age at maintenance as well as the path of investment between maturity and maintenance need to be optimized. However, senescence is still not an option. Any decline in size (i.e. an increase in mortality) is precluded by the state ratchet. To arrive ultimately at a framework where senescence is a possible optimal outcome the basic model has to be complicated even further.

4.4 An Optimization Model that Leads to Senescence

The state ratchet implies that any single-state life-history model along the general lines described above will always yield growth, declining mortality and increasing fertility followed by maintenance mode. Even if an exogenous event reduces ξ to some lower level ξ^-, then the individual would simply resume growth with the π-strategy previously followed at ξ^-.

In this kind of model, the single variable size ξ determines the capability of an individual to gather resources, to produce progeny and to avoid death. This spectrum might be too broad to be captured by size alone. Size can be measured by weight, length, number of cells, number of modular units or some similar index. While body size is determined by the number of cells and may remain constant, the functioning of cells may decline due to insufficient investment in maintenance because each cell is subject to continuous wear and tear. Therefore, it seems reasonable to distinguish between quantity and quality of cells. Functioning can be captured by a second state variable denoted by the Greek letter υ, which can take values between one and zero. The "vitality" of an individual can then be modeled as the product of ξ times υ, size weighted by functioning. Adding a second state variable to the model is a way to escape the state ratchet.

The model can be reformulated as follows. Fertility is given by

$$m(a) = \varphi\left(\pi(a) - \pi^2(a)\right) \xi(a)\, \upsilon(a)\,, \tag{4.42}$$

and mortality is given by

$$\mu(a) = \frac{b}{\xi(a)\, \upsilon(a)} + c\,. \tag{4.43}$$

Note that both fertility and mortality now depend on the product of size and functioning, $\xi(a)\,\upsilon(a)$, which captures vitality. The particular nonlinearity in fertility was retained.

This model can lead to determinate growth. Let a^* be the age at which growth is completed. Then $d\xi/da = 0$ for all $a > a^*$, where $\xi(a^*) = \xi^*$ denotes the size attained at the end of the determinate growth period. For $a < a^*$, functioning does not change, i.e. $\upsilon(a) = 1$. If investment falls below maintenance level, i.e. $\pi(a^*) < \delta_0 + \delta_1\, \xi(a^*)$ at a^*, functioning starts to deteriorate exponentially at the rate $\acute{\upsilon} = \kappa\,(\pi(a) - \delta_0 - \delta_1\, \xi^*)$ with initial condition $\upsilon(a^*) = 1$. If $\pi(a^*)$ is chosen to equal the deterioration at that age, the individual maintains its current functioning: this corresponds to the case of determinate growers with sufficient repair or replacement of tissues to escape senescence. The age a^* is not necessarily identical to age at reproductive maturity α, although for many determinate growers the two approximately coincide. The parameter combinations I used in the algorithm led to strategies for which $a^* = \alpha$.

Growth in ξ is positive until determinate size is attained and zero afterwards:

$$\frac{\frac{d\xi(a)}{da}}{\xi(a)} = \begin{cases} k\,(\pi(a) - \delta_0 - \delta_1\, \xi(a)) & \text{if } \pi(a) > \delta_0 + \delta_1\, \xi(a) \\ \\ 0 & \text{otherwise,} \end{cases} \tag{4.44}$$

where $\xi(0) = 1$. Functioning is constant at one until determinate size is reached and then declines:

$$\frac{\frac{dv(a)}{da}}{v(a)} = \begin{cases} 0 & \text{if } a < a^* \\ \kappa\,(\pi(a) - \delta_0 - \delta_1\,\xi^*) & \text{if } a \geq a^* \end{cases} \tag{4.45}$$

where $v(0) = 1$. Note that $\pi(a) - \delta_0 - \delta_1\,\xi^* < 1$. The parameters k and κ determine the speed of increase in size and the speed of decline in functioning, respectively.

Figure 4.4 exemplifies the optimal trajectories of $\pi(a)$, $\xi(a) \cdot v(a)$, $\mu(a)$ and $m(a)$ for determinate growth for this model. The results were obtained numerically. The maximum attainable size is $\xi = 25$; this size is almost reached at age of reproductive maturity α.

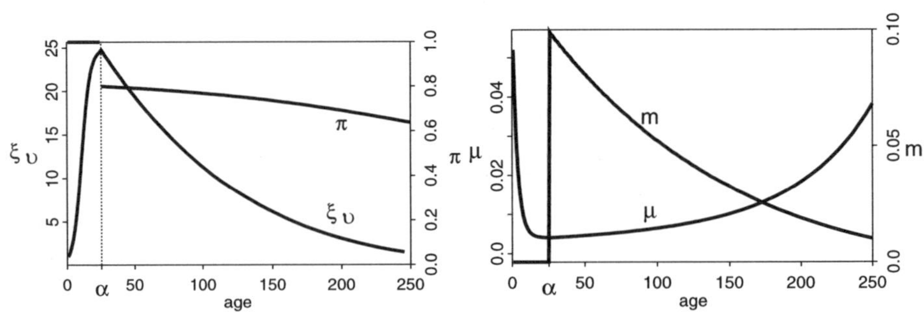

Fig. 4.4. $\xi(a){\cdot}v(a)$, force of mortality μ and fertility m resulting from optimal strategy $\pi(a)$ as a function of age a, for model with parameters $k = 3$, $\delta_0 = 0.9$, $\delta_1 = 0.004$, $\kappa = 0.05$, $b = 0.05$, $c = 0.002$, $\varphi = 0.02$

In this model, the state variable that effectively determines the strategy switches from size to functioning at age a^*. Before age a^* size is the only effective state variable, since functioning is constant. After age a^* functioning is the only effective state variable, since size is constant. Therefore, the state ratchet applies and functioning cannot increase again once it has fallen below one. The switch between size and functioning is assumed to occur only once. Growth cannot be resumed.

Another possibility for overcoming the state ratchet, but keeping a model that is essentially based on a single state, is to introduce a switch variable, which is a binary indicator that determines whether the organism is in up or down mode. The switch itself does not affect survival or reproduction. To jump the maintenance barrier, the switch needs to change from up into down mode. In this case the optimality of the strategy is not violated, as the smaller state is now associated with a different value of the switch. Depending on whether the switch is triggered once or several times, internally or externally, different state trajectories can emerge. Any repeated trajectories of increase and decrease have to be identical. This line of reasoning will be taken further in the next chapter.

4.5 Discussion

The first, simplest, model developed above led to sustenance as the only possible life-history strategy. The function describing reproduction had to be made nonlinear to get divergence from this prototype life history. The slightly more complex model led to a variety of possible life-history strategies between sustenance and enhancement. But senescence could still never be optimal.

To arrive at senescent strategies the state of the individual had to become more complicated, now being, effectively, a product of two variables, size and functioning. The product of size and functioning can be interpreted as reflecting the vitality of the individual. Vitality and not size determines mortality and fertility. Consequently it is possible that individuals might maintain about the same body weight, length or cell number over an extended period of life but suffer a decline in vitality due to wear and tear and lack of repair.

Although the eventuality was not considered here, size could increase over an extended period of life with this growth counterbalancing forces of deterioration and functional decline. In such species the ability to escape mortality, as captured by ξ times v, may remain roughly constant—resulting in non-senescence.

Note the distinction between senescence, on the one hand, and deterioration and functional decline, on the other. The term senescence is used only with regard to entire organisms, not parts of organisms. In this model deterioration is captured by $\delta(a)$ and decline in functioning by a decrease in $v(a)$. A tendency for existing body parts to deteriorate and to require repair or replacement to maintain functioning may possibly be a "fundamental, universal, and intrinsic" property of living organisms [9]; senescence, as defined here, is not.

The theoretical results of this chapter and the empirical evidence presented in Sect. 3.3.3, suggest the following hypotheses:

- Senescence characterizes individuals in species that attain a size at reproductive maturity that is close to maximum size. Such determinate- growth species include mammals, birds, insects and some other species including the nematode worm *C. elegans*. The

main model species studied by gerontologists are mammals (including humans, rats and mice), insects (especially *Drosophila* but also Medflies and some other insect species), *C. elegans*, and yeast. All of these species fall into this determinate- growth category. Many determinate- growth species also have fixed oocyte stocks or are otherwise limited with regard to reproductive capacity. Species that experience declines in fertility with age or that have limited fertility seem likely to suffer senescence.
- Non-senescence characterizes individuals in species that attain a size at reproductive maturity that is less than maximum size and that gain reproductive capacity as they grow. Such species with indeterminate growth include most trees, many other perennial plants, many modular animals such as corals and perhaps sponges, some kinds of algae, many fish, reptiles and amphibians, and probably various nonmodular invertebrates such as some mollusks and some echinoderms.

Species falling into the second category are not typically model organisms in gerontological research. This might be one reason why the universality of senescence was accepted as gerontological dogma.

Many biologists would agree that, for many species, stage is what determines mortality and fertility rather than age. If age itself matters at all, this line of thinking leads to the conjecture that biological age may be better captured by the "average age" of an individual — i.e., by some appropriate measure of the average age of the organs, body parts or cells of an individual — than by the chronological age of the individual. In indeterminate- growth species, continuing increases in size keep average age below chronological age. Furthermore, organisms that can repair, replace or rejuvenate body parts may show, over chronological time, slow increases or even decreases in average age. For instance, trees that replace their leaves annually, that develop new roots and new branches to replace damaged or lost ones, and that continue to grow may be of an average individual age that remains roughly constant and may even decline with chronological age. For some species of plants and animals, there can be a complete turnover of body parts over a time interval: for these species, average individual age can be much lower than chronological age and can decline over time if the individual grows and its component parts continue to renew themselves with time.

A remarkable example is Hydra [123]. Most species as small as hydra have a short life expectancy. Hamilton's reasoning would imply that hydra should senescence quickly after having lived past its typical lifespan in the wild. Contrary to this prediction, mortality is constant and has been effectively zero for hydras kept in the laboratory of Daniel Martinez for four years. Because there is rapid turnover of a hydra's cells, this example directs attention to considering not only size, i.e. quantity of cells, but also quality of cells. The first two models developed in this chapter consider size only, while the third model is a first attempt to incorporate not only quantity but also quality of cells. The model I develop in the following chapter accounts for both quantity and quality of cells.

This chapter has shown that non-senescence is a life-history strategy that is theoretically possible. Senescence can be avoided by "conceivable" organisms, namely by species with size-dependent vital rates. This finding together with the empirical evidence presented in Sect. 3.3 leads me to the hypothesis that non-senescence may indeed be a life history followed by some and maybe many plant and animal species. In the following chapter I develop a more general model to further study the evolution of senescence vs. non-senescence.

4.6 Next Steps

A critical examination of the model developed above indicates several directions to explore.

- The nonlinearity in fertility was introduced by means of efficiency of reproduction. Is there a more elegant way to incorporate efficiency?
- Reproduction and growth relate directly to size. This implicitly assumes that available resources are proportional to size. Is there a more realistic way to model resources?
- The vitality of an organism was modeled as a product of the two states size and functioning, in order to develop a model that can lead to non-senescent as well as senescent life-history strategies. The resulting model specifications seem rather complicated. Furthermore this model is not able to capture a simultaneous increase in size with a decrease in functioning. Size and deterioration were assumed to remain constant once functioning starts to decline. An idea for getting around this complication was suggested in Sect. 4.4.

The following chapter will take these points into account.

5

An Optimization Model Based on Vitality

The models developed in Chap. 4 show that non-senescence can be optimal. Size constitutes the central state variable in this framework. Mortality falls with increasing size and reproductive potential rises. The case of determinate growth, however, poses a challenge to this framework. Determinate growers, such as humans, often reach their final size at about the age of maturity. While size remains constant after the onset of reproduction, mortality steadily rises. This is incompatible with the strict size-dependence of mortality. A new model can be developed to address the deficiencies of the size-based model. To capture changing mortality at a constant size, the quality of size will be considered. The approach is rationalized in the following way. Even if size remains unchanged, all cells progressively accumulate damage over time and deteriorate. *Vitality*, defined as an individual's size adjusted for the functioning of body cells, can decline and therefore mortality can increase despite a constant body size. This notion was introduced in Sect. 4.4, where vitality was defined as the product of two functions, size and functioning. Here, vitality captures the accumulated functioning of all body cells, i.e. if a cell has been damaged and only works at 80 % of the capacity of an undamaged cell, this cell will account for 0.8 units of total vitality.

Facing ubiquitous decay, life is sustained by processes of regeneration and rejuvenation. The continuous creation of new, undamaged cells counterbalances deterioration. This balance determines whether or not vitality declines. The level of rejuvenation and repair depends on the trade-offs between reproduction on the one hand and growth and maintenance on the other. The optimal schedule of resource allocation determines the optimal trajectory of vitality. Increasing vitality raises reproductive potential and lowers mortality. Reproduction results in offspring but entails slower growth or even decline in vitality. The trajectory of vitality over age determines the age-trajectories of fertility, mortality and growth. The following evolutionary-demographic model sheds light on the fundamental questions of life-history theory based on the single state variable, vitality.

Anderson [8] developed a model based on the variable vitality. Anderson defines vitality as a randomly varying component of mortality which leads to death if vitality ever reaches zero. The use of the state variable vitality, as defined here, constitutes a new approach to life history modeling.

5.1 The Vitality Model

Survival is a function of mortality. In accordance with the size-based models it seems natural to model mortality as an inverse function of vitality, denoted by ψ. A simple function for the force of mortality, μ, is

$$\mu(\psi) = \frac{b}{\psi} + c\,, \tag{5.1}$$

where b and c are constant parameters. The intrinsic parameter b captures all causes of death an individual can escape from by increasing its vitality, while the extrinsic parameter c captures the always prevalent, non-zero risk of death. Note that "extrinsic" and "intrinsic" refer to vitality-dependent vs. vitality-independent mortality.

Reproduction and growth depend on the level of available energy. In the size-based models, energy was simply proportional to size. However, energy production is not equivalent to size but has been found to scale allometrically with it [107]. A sound theoretical basis for a particular relation between size and net energy available was given by West et al. [210], their Equation (3). This formula captures the difference between energy created by cell metabolism and energy required for it, based on an allometric relation between size and energy production.

The model developed in this chapter uses Equation (3) from West et al. [210] to determine the available resources of an individual at its current level of vitality. The formula of West and colleagues [210] is based on the variable size. The link between vitality and size is assumed to be tight enough to justify the substitution of vitality for size in this equation for this specific model. Net energy production, denoted by $\epsilon(\psi)$, depends on the difference between build-up and break-down processes at current vitality,

$$\epsilon(\psi) = k\,\psi^{0.75} - \kappa\,\psi\,, \tag{5.2}$$

where k and κ are constant parameters. Anabolic, build-up processes are directly linked to metabolic output, which is assumed to be proportional to vitality to the power 0.75. Catabolic, break-down processes are assumed to be proportional to vitality to the power one.

The exact value 0.75 for anabolic processes was thought to be a so called life-history invariant [33]. The method of calculating these life-history invariants has recently been called into question [49, 137]. The particular value of 0.75 might therefore not be invariant across species. The qualitative results of my model, however, do not depend on the particular value 0.75 but only require the existence of such an allometric relation.

Energy production is maximal at vitality ψ_ϵ

$$\psi_\epsilon = \left(\frac{3}{4}\frac{k}{\kappa}\right)^4 . \tag{5.3}$$

As in the size-based model, growth and maintenance are paid out of the same budget. Part of the energy available must be used to offset the declining functioning of cells. The change in vitality is given by the difference between the fraction of resources allocated to growth (newly built cells) and the unavoidable deterioration of functioning of current cells at a constant rate δ. Damage is proportional to vitality and integrates naturally into the structure of West et al.'s equation. Consequently, vitality ψ changes over time according to

$$\dot{\psi} = \pi(\psi)^{\eta_g}\,\epsilon(\psi) - \delta\,\psi\,, \tag{5.4}$$

where $\pi(\psi)$ denotes the fraction of energy allocated to growth, as in the models in Chap. 4. In contrast to those models, $\pi(\psi)$ can now have a nonlinear effect on the change in state depending on the value of the constant parameter η_g (g for growth). In the extreme case of no energy allocation to growth and maintenance, vitality deteriorates exponentially and, as in the size-based model, mortality rises exponentially. The reasoning behind the incorporation of parameter η_g will be given below.

The level of initial vitality is $\psi(0) = 1$ and initial time zero corresponds to time at birth[1]. Vitality is treated as a dimensionless variable,

[1] Note that the model in its current form does not account for stage-specific life-histories.

assuming that vitality is normalized by dividing through with a reasonable base unit. If the state of an organism at birth corresponds to some $\psi_{real}(0)$, this implies that vitality $\psi(a)$ in this model is given by[2]

$$\psi(a) = \psi_{real}(a)/\psi_{real}(0) \tag{5.5}$$

and therefore

$$\psi(0) = \frac{\psi_{real}(0)}{\psi_{real}(0)} = 1\,. \tag{5.6}$$

The level of $\pi(\psi)$ that corresponds to maintenance of current vitality can be derived from (5.4). Denoting the level of $\pi(\psi)$ at $\dot{\psi} = 0$ by π_0 and inserting (5.2) yields

$$\pi_0 = \left(\frac{\delta}{k\,\psi^{-0.25} - \kappa}\right)^{\frac{1}{\eta_g}}\,. \tag{5.7}$$

Vitality cannot increase indefinitely. An upper limit to ψ, denoted by Ψ, is reached at maximum investment $\pi(\psi) = 1$ and $\dot{\psi} = 0$,

$$\Psi \equiv \left(\frac{k}{\kappa + \delta}\right)^4\,. \tag{5.8}$$

Available energy must be nonnegative. This implies that

$$\psi \leq \left(\frac{k}{\kappa}\right)^4 \tag{5.9}$$

must hold. This is always true since (5.9) implies that ψ cannot exceed maximum attainable vitality Ψ, as given by (5.8).

In the initial size-based model (Sect. 4.3) reproductive effort and reproductive output are related linearly. As explained in Sect. 4.3.2, it turns out that this assumption restricts optimal solutions to energy allocation exclusively to either growth or reproduction. To develop a model that permits a broad scope of possible investment strategies, a nonlinear influence of investment needs to be incorporated that still includes the possibility of exclusive allocation. This is the technical argument that motivates the introduction of parameter η_g in (5.4). The biological motivation for introducing nonlinear effects is the following.

[2] If functioning at birth is assumed to be perfect, then $\psi_{real}(0)$ is equal to the number of cells (corresponding to the minimum size) at birth. In order to establish the real vitality scale ψ_{real} from the algorithm, vitality has to be multiplied by $\psi_{real}(0)$.

Growing a human arm requires considerable effort and is so difficult that, if the arm is lost, no new arm can regrow. In contrast, growing a branch of a tree can be done readily to increase size or replace broken branches. The growth apparatus in humans and trees is inherently different. In the former case, it might be very costly and even impossible to keep or rebuild the machinery that would allow the regrowth of a lost arm. In the latter case, maintenance is cheap because existing machinery can be used to maintain the organism without much additional cost.

Parameter η_g captures the nature of the growth and maintenance apparatus of a species. When η_g exceeds one, the investment function π^{η_g} in (5.4) is convex. The marginal benefits in outcome become larger as π approaches one. Note that the convexity favors exclusive investment strategies. When η_g is below one, the investment function π^{η_g} is concave. The marginal benefits in output become smaller as investment approaches one. Note that concavity favors intermediate investment strategies. The parameter η_g in (5.4) captures the returns to scale in growth and maintenance investment. The parameter can also be interpreted as the efficiency of the growth system. Values of η_g below one correspond to efficient, i.e. cheap, growth, and values of η_g above one correspond to inefficient, i.e. costly, growth.

Figure 5.1 illustrates the influence of parameter η_g via investment π on the change in vitality. Note that the change in vitality is always larger for a given level of investment π when η_g is below one as opposed to being above one. Likewise, any particular level of change in vitality requires a smaller investment, given that η_g is below one rather than above one. Note further that values of η_g below one imply a concave shape, while values above one correspond to a convex shape of the change in vitality with increasing investment.

In the modified size-based model (Sect. 4.4) an arbitrary attempt was made to introduce nonlinearity with respect to reproductive effort. In the vitality model the maternity function is specified as

$$m(\psi) = \varphi\,(1 - \pi(\psi))^{\eta_r}\,\epsilon(\psi)\,. \tag{5.10}$$

In accordance with the size-based models, fertility is proportional to available energy, in this model $\epsilon(\psi)$, and reproductive effort, $1 - \pi(\psi)$. In contrast to the size-based model, nonlinearity in reproductive effort is incorporated by parameter η_r (r for reproduction) which captures the efficiency of reproduction, analogous to η_g. As in the size-based models, the constant φ is a scaling parameter set to the value that ensures that optimal lifetime reproduction is equal to one and, hence, $r_{max} = 0$.

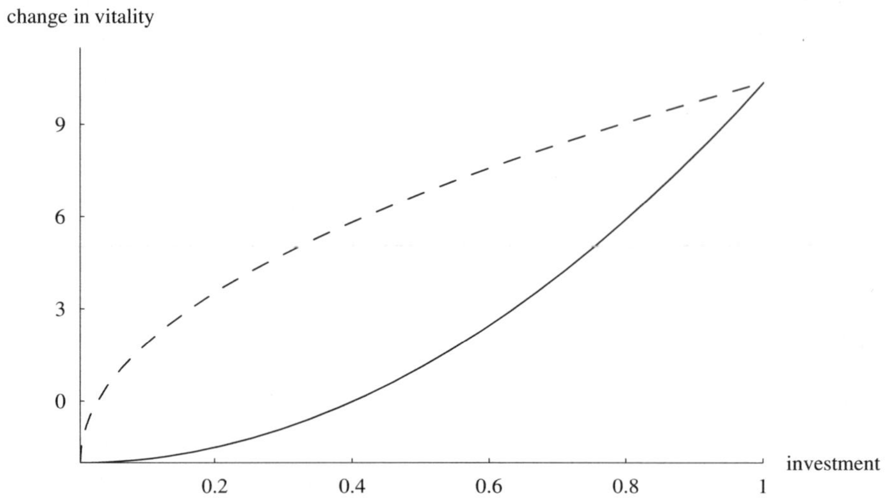

Fig. 5.1. The influence of η_g via investment π on the change in vitality as specified by (5.4) (*The dashed line exemplifies values of η_g below one, in particular $\eta_g = 0.5$. The solid line exemplifies values of η_g above one, in particular $\eta_g = 2$. In both cases $\psi = 20$, $k = 3$, $\kappa = 0.8$ and $\delta = 0.1$.*)

The manner in which nonlinearities enter the model is biologically and technically motivated. The approach makes use of the well-known concept used in economics of the Cobb Douglas production function. Each input factor to the production function is raised to a power reflecting how efficient each factor, in economics labor and capital, is in producing output. Two new parameters (that influence the optimal trajectory of investment) enter the model as exponents of investments. Power functions have previously been used to introduce nonlinearities into life-history models [26, 37, 39, 62, 175]; see Charlesworth [27, Sect. 5.3.4.] for review). In particular, the importance of the shape of the investment function for the optimal life history strategy has been recognized. In their reproductive effort models, Gadgil and Bossert [62] and Schaffer [175] found that concave investment functions favor iteroparous strategies (repeated breeding, i.e. intermediate reproductive effort) while convex investment functions favor semelparous strategies (a single breeding event, in which reproduction is fatal, i.e. exclusive investment).

George E. P. Box said: “All models are wrong, but some are useful.” [13] Models are wrong because they simplify the complexity of life. But without this simplification, patterns can hardly be observed and understood. A useful model captures the most important aspects of reality, reveals general patterns and provides a source for hypotheses that could explain basic processes of life. Such a model, although necessarily wrong, enhances our understanding of nature.

Adding efficiency to the size-based model increases complexity but it also considerably broadens the model's potential for predicting various life-history strategies. The non-linearities capture cases in nature when parallel investment in growth and reproduction is optimal. Therefore, these extensions to the model can be justified as a useful complication to a still simple model.

5.1.1 The Parameters

k, κ and δ

Parameter k captures the speed of growth of vitality (see (5.2) and (5.4)). Faster growth implies a rapid fall in mortality (5.1) and reduces the time of development. Furthermore, higher values of k decrease maintenance costs (5.7) and increase maximum vitality (5.8). Parameter κ is inversely related to maximum vitality. Elevating κ slows growth, increases maintenance costs (5.7) and decreases maximum vitality (5.8). Parameter δ determines the rate of decline in vitality (5.4). Higher δ increases maintenance costs (5.7) and decreases maximum vitality (5.8).

If all available energy is allocated to reproduction, then δ determines the constant rate of increase in mortality (5.1). A decline in vitality implies not only a reduction in survival but also in reproductive potential. Therefore, larger values of δ will tend to increase the investment of resources in growth in order to slow down the deterioration process.

Parameters k and κ determine the shape of the energy trajectory over vitality (5.2). If $\kappa < 3\delta$, then energy is an increasing function of vitality because the maximum attainable vitality is smaller than the level of vitality that maximizes energy, $\Psi < \psi_\epsilon$. Otherwise, if $\kappa > 3\delta$, then the trajectory of energy is hump-shaped with respect to vitality. The influence of the relation between κ and δ on the energy trajectory over vitality is visualized in Fig. 5.2. Note that an increase in vitality beyond the threshold given by (5.3), which corresponds to the peak of energy, can only be optimal if the corresponding reduction in mortality offsets the loss in available resources, i.e. in growth and reproductive potential.

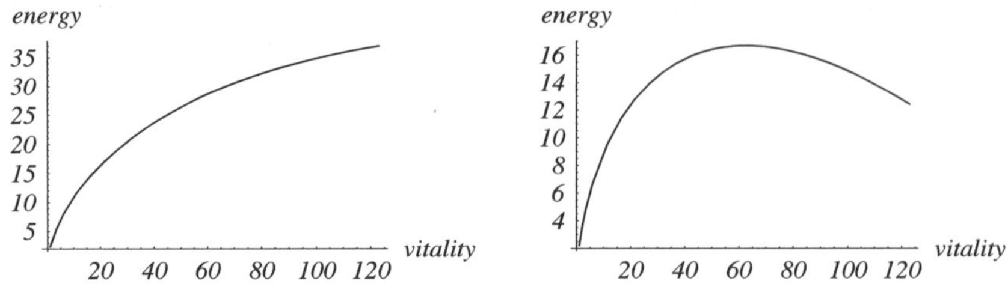

Fig. 5.2. Comparison of trajectories of energy over vitality for two parameter combinations that lead to a maximum attainable vitality of $\Psi = 123$ but imply different shapes (*left:* $k = 3$, $\kappa = 0.6$, $\delta = 0.3$; *right:* $k = 3$, $\kappa = 0.8$, $\delta = 0.1$)

The parameters k, κ, and δ set the speed of growth and decay and can therefore be used to determine the time and size scale of the strategy. Getting a handle on measurable quantities like time and size in this model is one future project that naturally follows from my work (see Chap. 6).

b and c

Parameters b and c determine the overall level of mortality (5.1). Parameter b captures the state-dependent, intrinsic component of mortality, i.e. b determines how important it is to attain and maintain a high level of vitality. Reasonable magnitudes of parameter b are given by the fact that $b/\psi(0)$ determines infant mortality. Furthermore, the minimum level of state-dependent mortality depends on parameter b and on maximum vitality Ψ and is given by b/Ψ. Parameter c captures the state-independent, extrinsic mortality component. The overall level of infant mortality is given by $b/\psi(0) + c$ and the minimum mortality that can be attained is given by $b/\Psi + c$.

The influence of extrinsic and intrinsic mortality in this model is investigated below (see Sect. 5.5.3).

η_r and η_g

Parameter η_r captures the intrinsic costs of reproduction (5.10). It determines the propensity to share resources between reproduction and growth. Clearly, if an organism follows an exclusive strategy, i.e. either reproduction or growth and repair, then π equals one or zero and an exponent will have no influence. However, if energy is shared between processes, then larger values of η_r reduce the reproductive output that could have been achieved with the same level of investment at lower values of η_r. Values below one favor parallel investment in growth and reproduction.

Parameter η_g captures the intrinsic costs of growth and determines the maintenance costs of a certain level of vitality (5.7). A large value of η_g implies higher maintenance costs at each level of vitality. Therefore, low values of η_g favor non-senescence strategies. During periods of parallel growth and reproduction, higher η_g implies a reduced speed of growth.

Both parameters η_r and η_g capture the efficiency of energy use and determine how advantageous it is to specialize in growth and reproduction, i.e. how costly it is to run a growth and reproduction system in parallel. The costs of reproduction and maintenance are expected to crucially determine the optimal energy allocation between reproduction and growth. In this chapter I will investigate whether or not this expectation is fulfilled.

5.2 The Vitality Model as a Control Problem

The model developed in the previous section is an autonomous control problem with an infinite time horizon. In the following sections the problem is formulated and subsequently a solution is approached including a discussion of the range of possible optimal solutions.

5.2.1 Problem Formulation

The objective function to be maximized is given by

$$\max_{\pi} \int_0^{\infty} e^{-\phi}\,(1-\pi)^{\eta_r}\,\epsilon(\psi)\,da \tag{5.11}$$

where

$$\epsilon(\psi) = k\,\psi^{0.75} - \kappa\,\psi \tag{5.12}$$

is associated with the level of energy available, as defined earlier. The cumulative hazard of death, ϕ, is defined as the logarithm of survival $l(a)$ at age a

$$\phi(a) = -\ln\left(l(a)\right)\,. \tag{5.13}$$

The only control variable in this problem is the proportion of investment $\pi(a) \in [0,1]$ towards growth and survival versus reproduction. The state variables of this problem are vitality $\psi(a) \in \mathbf{R}^+$ and the cumulative hazard of death $\phi(a) \in \mathbf{R}^+$.

The change in vitality with age is given by

$$\dot{\psi} = f(\psi, \pi) \equiv \pi^{\eta_g} \, \epsilon(\psi) - \delta \, \psi \,, \tag{5.14}$$

and the change in the cumulative hazard of death is given by

$$\dot{\phi} = \mu(\psi) \,, \tag{5.15}$$

obeying the initial conditions[3]

$$\psi(0) = 1 \tag{5.16}$$

and

$$\phi(0) = 0 \,. \tag{5.17}$$

The Hamiltonian function[4] associated with this problem is given by

$$H = e^{-\phi} \, (1 - \pi)^{\eta_r} \, \epsilon(\psi) + \lambda_\psi \, \dot{\psi} + \lambda_\phi \, \dot{\phi} \,, \tag{5.18}$$

i.e.

$$H = e^{-\phi} \, (1 - \pi)^{\eta_r} \, \epsilon(\psi) + \lambda_\psi \, (\pi^{\eta_g} \, \epsilon(\psi) - \delta \, \psi) + \lambda_\phi \, \mu(\psi) \,, \tag{5.19}$$

with the transversality conditions for the two co-state variables[5]

$$\lambda_\psi(\infty) = \lambda_\phi(\infty) = 0 \,. \tag{5.20}$$

Note that the Hamiltonian function, denoted simply as H, is a function of the control, state and costate variables but does not explicitly depend on time t, i.e. $H = H(\pi(t), \psi(t), \phi(t), \lambda_\psi(t), \lambda_\phi(t))$.

[3] Note that the model does not account for optimization of size at birth. This is an interesting topic that could be explored with an extended version of this model. Including variable size at birth which can be reasonably interpreted as vitality at birth implies several issues that will be discussed in a future joint paper by Kenneth Wachter and me.

[4] see Sec. 4.3.2

[5] Note that the subscripts 'ψ' and 'ϕ' to λ should not be confused with denoting partial derivatives.

5.2.2 Solution

The Maximum Principle requires that an optimal control path has to maximize the Hamiltonian function H. If H is differentiable w.r.t. π the optimal π^* (star indicates "optimal") can be found by

$$\frac{dH}{d\pi} = -\eta_r\, e^{-\phi}\,(1-\pi)^{\eta_r-1}\,\epsilon(\psi) + \eta_g\,\lambda_\psi\,\pi^{\eta_g-1}\,\epsilon(\psi) = 0\,. \tag{5.21}$$

Rearranging this expression yields

$$\frac{(1-\pi)^{\eta_r-1}}{\pi^{\eta_g-1}} = \frac{\eta_g}{\eta_r}\,e^{\phi}\,\lambda_\psi\,. \tag{5.22}$$

A maximum further requires the second derivative of H to be negative, and therefore condition

$$\begin{aligned} H_{\pi\pi} &= \eta_r\,(\eta_r-1)e^{-\phi}\,(1-\pi)^{\eta_r-2}\,\epsilon(\psi) \\ &\quad + \eta_g\,(\eta_g-1)\lambda_\psi\,\pi^{\eta_g-2}\,\epsilon(\psi) < 0 \end{aligned} \tag{5.23}$$

must hold, which after rearranging becomes[6]

$$(\eta_r-1)\,\frac{(1-\pi)^{\eta_r-2}}{\pi^{\eta_g-2}} < -(\eta_g-1)\,\frac{\eta_g}{\eta_r}\,e^{\phi}\,\lambda_\psi\,. \tag{5.24}$$

If both conditions (5.22) and (5.24) are true, expression (5.22) can be substituted for in (5.24), yielding

$$(\eta_r-1)\,\frac{(1-\pi)^{\eta_r-2}}{\pi^{\eta_g-2}} < -(\eta_g-1)\,\frac{(1-\pi)^{\eta_r-1}}{\pi^{\eta_g-1}}\,, \tag{5.25}$$

and after rearranging

$$(\eta_r-1)\,\frac{\pi}{1-\pi} < -(\eta_g-1)\,. \tag{5.26}$$

It should be emphasized that conditions (5.22) and (5.24) are not necessary for an optimal solution to exist. But if they are satisfied, then the sufficiency condition derived by Mangasarian [117] is satisfied. The Hamiltonian function is concave in π and an interior solution is optimal.

A Hamiltonian function that is linear (as in Sect. 4.3.2) or convex in π implies that a potential maximum can only be achieved at the boundaries of the feasible set of π. Thus this maximum cannot be

---

[6] I am not dividing by $\eta-1$ terms, since they could become negative depending on the parameter values in which case the inequality would turn around.

found by differentiating the Hamiltonian function as done in 5.21. The case of a convex Hamiltonian will be discussed later in this chapter.[7]

Condition (5.26) helps determine the range of parameters η_r and η_g for which it is safe to say that problem (5.11) has an optimal solution:

- It is immediately apparent that condition (5.26) is true for values of $\eta_r < 1$ & $\eta_g < 1$, $\eta_r = 1$ & $\eta_g < 1$, and $\eta_r < 1$ & $\eta_g = 1$. This range of the parameter space corresponds to a concave Hamiltonian function.
- If $\eta_r = 1$ & $\eta_g = 1$, the condition is violated. Instead, the Hamiltonian is linear in π and, as discussed in Sect. 4.3.2, the optimal solution is singular.
- For values $\eta_r > 1$ & $\eta_g < 1$ and $\eta_r < 1$ & $\eta_g > 1$ the Hamiltonian can be concave as well as convex, depending on the value of π and on the relative magnitude of η_r and η_g. I will further investigate this very interesting case later in this chapter.
- If both $\eta_r > 1$ & $\eta_g > 1$ the concavity condition is violated, the Hamiltonian is convex. Do optimal solution exists for problem (5.11)? And if not, is there a way to modify the model in order to get optimal solutions for the case of $\eta_r > 1$ & $\eta_g > 1$? As for the previous case I will tackle those questions later in this chapter.

Maximizing the Hamiltonian function is not the only condition the Maximum Principle requires to be fulfilled. As discussed in Sec. 4.3.2 in Chap. 4, the co-state variables have to meet the following conditions:

$$\dot{\lambda}_\psi = -\frac{dH}{d\psi} \tag{5.27}$$

and

$$\dot{\lambda}_\phi = -\frac{dH}{d\phi} \,. \tag{5.28}$$

Solving the differential equations (see App. A.1) yields the following results:

The shadow price of vitality at age a is given by the associated cumulated changes in fertility and mortality over all remaining ages discounted by the corresponding cumulative changes in growth.

$$\begin{aligned} \lambda_\psi(a) = \int_a^\infty & \left(e^{-\phi} \, (1-\pi)^{\eta_r} \, \epsilon_\psi + \lambda_\phi \, \mu_\psi \right) \\ & \times e^{\int_a^x \pi^{\eta_g} \, \epsilon_\psi \, - \, \delta \, ds} \, dx \,. \end{aligned} \tag{5.29}$$

[7] The existence of optimal solutions for my problem will be discussed in more detail in a forthcoming paper by Kenneth Wachter and me.

The shadow price of the cumulative hazard of death at age a is the negative value of remaining reproduction at age a, i.e. the penalty for having one unit higher cumulative hazard:

$$\lambda_\phi(a) = -\int_a^\infty e^{-\phi}\,(1-\pi)^{\eta_r}\,\epsilon(\psi)\,dx\,. \tag{5.30}$$

Equation (5.30) can be substituted in (5.29) to yield the final expression for $\lambda_\psi(a)$, being

$$\begin{aligned}\lambda_\psi(a) = {} & \int_a^\infty e^{\int_a^x \pi^{\eta_g}\epsilon_\psi - \delta\,ds} \\ & \times \left(e^{-\phi}\,(1-\pi)^{\eta_r}\,\epsilon_\psi + \tfrac{b}{\psi^2}\int_x^\infty e^{-\phi}\,(1-\pi)^{\eta_r}\,\epsilon(\psi)\,d\tau\right)dx\,.\end{aligned} \tag{5.31}$$

The shadow price of vitality is given by the benefits of increasing reproduction due to higher vitality as well as the gains in remaining reproduction due to lower mortality, both weighted by the change in growth. As long as an increase in vitality leads to faster growth, this weight is above one (revaluating), if the increase in vitality leads to slower growth, then the weight is below one (devaluating).

5.2.3 The Role of the Second State Variable

A condition for optimal investment is given in (5.22). This condition requires that $\lambda_\psi(a)$, given in (5.31), is multiplied by $\exp(\phi(a))$. It becomes apparent that an optimal solution depends only on the current value[8] $\lambda_\psi^c \equiv \exp(\phi(a))\,\lambda_\psi(a)$ which is not discounted by death up to age a, in other words the cumulative hazard of death from birth to age a is erased. Consequently, expression (5.22) and thus an optimal investment path after age a is independent of the state variable $\phi(a)$; the hazard of death accumulated between age zero and age a has no effect.

Why is that so? Why does the second state variable not influence the optimal solution? The answer can be found reformulating the control problem. Maximizing the objective function in (5.11) from age zero to infinity requires maximizing the objective from any time point T onwards. Thus, the objective can be written as

$$\max_\pi\left(\int_0^T e^{-\phi(x)}\,m(x)\,dx + \max_\pi\int_T^\infty e^{-\phi(x)}\,m(x)\,dx\right), \tag{5.32}$$

[8] The concept of current values, i.e. values at time t rather than their equivalent at time zero is well described in [90, pp. 164–174]

where $m(x)$ denotes the maternity function at age x. Maximization before age T is conditional on the maximization after T. But the maximization after T is conditional only on the value of $\psi(T)$. In particular the cumulative hazard of death between age zero and T does not influence the optimal strategy after age T. Instead, the factor $\exp -\phi(T)$ can be drawn outside the integral, i.e.

$$\max_{\pi} \int_T^{\infty} e^{-\phi}\, m(x)\, dx \;=\; e^{-\phi(T)} \max_{\pi} \int_T^{\infty} e^{-\phi(x-T)}\, m(x)\, dx \,, \tag{5.33}$$

revealing its nature as a mere scaling factor. Effectively, the cumulative hazard starts off at $\phi = 0$ for any time point the control problem is supposed to be solved from. It has no effect on the first state variable since the change in vitality is entirely driven by investment π and by vitality itself. The fact that the change in cumulative hazard ϕ depends on mortality which in turn depends on vitality further emphasizes the point that vitality is the only actual state variable that matters.

Thus, the control problem formulated in this chapter is essentially a single state, single control, autonomous, infinite horizon optimal control problem. As discussed earlier in Chap. 4 this implies that any optimal state path has to be monotone. Since life starts by growth, only initially increasing vitality trajectories are sensible. This means mortality cannot increase; senescence is impossible. Only growth followed by either a period of parallel growth and reproduction and then maintenance or development followed by maintenance directly are the possible strategies. Given the idea of "inevitable senescence" it is remarkable how challenging it is to actually develop a model that can lead to senescence as an optimal life history strategy.

In the size chapter I was able to modify my model by adding a second state variable to come up with solutions that yielded senescence. How could I change my vitality model to broaden the scope of possible solutions, including the pattern of human senescence? As mentioned above there are several parameter combinations for η_r and η_g for which the Hamiltonian function is convex. Maybe the "weird" cases of $\eta_r > 1$ & $\eta_g < 1$, $\eta_r < 1$ & $\eta_g > 1$ and $\eta_r > 1$ & $\eta_g > 1$ provide a rich ground for exploring exotic strategies? Before I make the leap to modifying my model let me step back for a moment and have another look at the results.

5.2.4 Hamilton and Reproductive Value - Revisited

First note that the shadow price of the cumulative hazard of death at age a given in (5.30) corresponds to the negative of remaining reproduction at that age, which is equivalent to the numerator in Hamilton's indicator for the force of selection on age-specific mortality. As discussed at length in Chap. 2, this quantity inevitably declines with age. But in my control problem it does not automatically mean that mortality becomes less important, and that mortality would ultimately increase with age. My results just showed that when the Hamiltonian function is concave, then vitality has to increase and thus mortality has to decline, despite an inevitable falling shadow price of the cumulative hazard of death.

Hamilton emphasized how important it was to use 'remaining reproduction' and not 'reproductive value' for quantifying selection pressure. Interestingly, reproductive value can be recovered if I modify my model formulation, though leaving it essentially unchanged: as just shown the cumulative hazard of death at age a has no influence on the optimal solution at age a. Looking closer at the problem formulation, the definition of the cumulative hazard $\phi(a)$ as the second state variable seems somewhat arbitrary. Indeed, I could have equally well let survival $l(a)$ serve as the second state variable. This would have given me a equation of motion, different to the one in (5.15), namely $\dot{l}(a) = -\mu(a)\, l(a)$, but again the condition for optimal π would have been independent of the second state. Doing the corresponding modifications and calculations leads to a shadow price of survival that is equal to

$$\lambda_l(a) = \frac{1}{l(a)} \int_a^\infty l(x)\,(1-\pi)^{\eta_r}\, \epsilon(\psi)\, dx \,. \tag{5.34}$$

This expression corresponds to the reproductive value at age a. Contrary to the shadow price of the cumulative hazard of death which inevitably declines with age, the shadow price of survival can also be constant or increase with age.

Both the quantities – remaining reproduction and reproductive value – that were of central importance to Hamilton's discussion of the evolution of senescence emerge as part of my life history optimal control problem. Furthermore, the quantities have been associated with opposite answers to the essential question of whether senescence is inevitable or not. Now it turns out that both quantities have their interpretation as shadow prices for either survival or cumulative hazard. They are weights in the Hamiltonian function. But neither of them actually influences the optimal control path (see (5.22)). The optimal control path only depends on the current value of the shadow price of vitality and the magnitude of the η parameters.

This is an important finding because it is tempting to predict the shape of mortality and fertility from the fitness sensitivities directly.

My result underlines the fact that one should not do so. Explaining the change in a trait (and ultimately the evolution of age-trajectories of mortality and fertility) requires multiplying the sensitivity in fitness by the variance-covariance matrix [105], as discussed in Chap. 3. Lande's so called G-matrix essentially contains all the trade offs among the fitness relevant traits.

People studying the evolution of senescence try to understand whether optimization approaches that rest on trade-offs or approaches based on fitness sensitivities (which is what Hamilton's indicators are and reproductive value is the fitness sensitivity with respect to survival) are a better way of explaining the evolution of senescence (see Chap. 3). I find that in my optimization framework fitness sensitivities appear as weights in the optimization formulas. Thus, both approaches are intertwined. The justification and importance for fitness sensitivities in shaping age-trajectories of mortality and fertility is in giving appropriate weights to the trade-offs that are balanced by evolution. Giving weights to trade-offs is exactly the same function that sensitivities take on in determining short term evolutionary change, when fitness sensitivities are multiplied by the appropriate G-matrix. For both approaches the central role of trade offs is conspicuous. Nailing down the trade offs - however - is one of the hardest nuts to crack for life history biologists, independently of which approach is taken.

5.3 The Constrained Vitality Model

Let me now get to the 'weird' ranges of parameters η_r and η_g. For values of η_r and η_g that lead to a convex Hamiltonian function no simple answer can be given as to whether an optimal solution exists or not. Mangasarian's theorem [117] is sufficient for an optimal solution but not necessary. An example will help to understand why convexity is a problem.

Let me focus on the case of $\eta_r > 1 \,\&\, \eta_g > 1$. If both η_r and η_g exceed one, then the Hamiltonian is convex in π. Therefore the highest value of H is found at the boundaries of the feasible set for values of π. Thus the only two possible values of π that could maximize H are either $\pi = 1$ or $\pi = 0$.

If $\pi = 1$ no reproduction is realized at all, so the Hamiltonian equals zero. If $\pi = 0$, then all resources are spend on reproduction, and vitality declines. For $\pi = 0$ the Hamiltonian function takes on some positive value, hence the Hamiltonian is larger for $\pi = 0$ than for $\pi = 1$, i.e. $H(\pi = 0) > H(\pi = 1)$. As $\pi = 1$ implies growth in vitality whereas

$\pi = 0$ implies shrinkage in vitality, π would have to remain at the level of zero forever because an optimal trajectory for this problem has to be monotone. It can be proven, however, that $H(0) > H(1)$ will be violated at some point, at least for the special case of constant mortality, $\mu = c$ (see App. A.2). This case illustrates the fact that the model has no optimal solution if both η's are larger than one[9].

An interesting aspect of the proof, however, is the insight that at some point the Hamiltonian will lead to higher values for $\pi = 1$ than for $\pi = 0$. Thus the strategy switches from full reproduction to full growth. One can show that the convex Hamiltonian produces zig-zagging strategies. And it is this zig-zagging that holds the key to the explanation for the non-existence of a solution[10]. To prevent this behavior one could constrain the vitality model in a way that zig-zagging is not an option. Let me explore this avenue further.

The mode of change of an organism can be constrained such that it can be changed only once. Hence, periods of growth and shrinkage cannot occur repeatedly but can only alternate once. Since life starts with growth, initially the mode of change for any organism is to increase in vitality. The organism is free to grow and increase in vitality until eventually maintenance level is reached or the organism could switch to shrinkage. Once the organism is on a decreasing vitality path, it will eventually reach maintenance but cannot get back on an increasing path. For the example discussed above this would imply that initially π equals one, at the onset of reproduction π switches to zero and at the age when λ_ψ rises above one, π rises up to its maximum permissable level, i.e. maintenance $\pi = \pi_0$. In this way, the problem encountered without the constraint disappears. There is an optimal solution inside the range of feasible trajectories of π. It is clear that this strategy could be beaten in the unconstrained model, but for the modified, constrained version π_0 is the best feasible strategy.[11]

Let me formalize the constrained model. I will introduce a second control variable defined as the time point $T \in [0, \infty]$ of switch between growth and shrinkage. Since life starts with growth, all ages younger than T are associated with increasing vitality. Note that theoretically, optimal strategies can imply pure shrinkage ($T = 0$; initially being in shrinkage mode) as well as pure growth strategies ($T = \infty$; never switch

---

[9] As discussed in more detail in a forthcoming paper by Kenneth Wachter and me.

[10] Ditto.

[11] It may well be that for some species it is physiologically impossible to switch from a senescent back to a non-senescent trajectory, i.e. from increasing back to decreasing mortality, and thus my constraint might not be unrealistic.

to shrinkage). Thus, the change in vitality is given by

$$\dot{\psi} = max[f(\psi, \pi), 0] \text{ for } a \text{ in } [0, T) , \tag{5.35}$$

where f is the same as in the unconstrained model given in (5.14). Equation (5.35) implies that any feasible strategy π associated with negative changes in vitality in the unconstrained model is now mapped on the zero-line for all ages before T. After T the change in vitality is given by

$$\dot{\psi} = min[f(\psi, \pi), 0] \text{ for } a \text{ in } [T, \infty] . \tag{5.36}$$

Now the objective function can be written as

$$\max_{\pi, T} \left(\int_0^T l(x)\, m(x)\, dx + e^{-\phi(T)} \int_T^\infty e^{-(\phi(x)-\phi(T))}\, m(x)\, dx \right) , \tag{5.37}$$

using general notation for survival $l(x)$ and reproduction $m(x)$. Applying (5.33) this expression is equivalent to

$$\max_{\pi, T} \left(\int_0^T l(x)\, m(x)\, dx + e^{-\phi(T)} \max_{\pi} \int_T^\infty e^{-(\phi(x)-\phi(T))}\, m(x)\, dx \right) . \tag{5.38}$$

It turns out that the constrained model consists of two optimization problems that are linked. The inner maximization problem in (5.38) is given by

$$\max_{\pi} \int_T^\infty e^{-(\phi(x)-\phi(T))}\, m(x)\, dx . \tag{5.39}$$

Note that the mode of change in vitality for the inner problem is shrinkage. The initial conditions for the inner problem are

$$\psi^S(T) = \psi^G(T) \tag{5.40}$$

and

$$\phi^S(T) = \phi^G(T) . \tag{5.41}$$

The upper case letters S and G indicate the different modes of change for the inner and the outer problem. The outer problem (where vitality can not decrease) sets initial conditions for the inner problem (where vitality can not increase) but the inner problem can otherwise be solved independently of the outer one.

The change in vitality for the inner problem is given by (5.36) and the transversality conditions are as before

$$\lambda_{\psi}^{S}(\infty) = \lambda_{\phi}^{S}(\infty) = 0 \,. \tag{5.42}$$

The Hamiltonian function is essentially the same as in the unconstrained model, and the solutions for λ_{ψ}, λ_{ϕ} and ψ remain the same, just with new initial vitality level $\psi(T)$. An optimal solution for this problem again has to be monotone, therefore vitality can only decline until zero or some level that is maintained.

The outer maximization problem depends on the maximum value of the objective of the inner problem. The initial conditions of the outer problem are

$$\psi^{G}(0) = 1 \tag{5.43}$$

and

$$\phi^{G}(0) = 0 \tag{5.44}$$

while now the transversality conditions are unusual in the way that they connect the outer problem to the inner problem,

$$\lambda_{\psi}^{G}(T) = e^{-\phi(T)} \lambda_{\psi}^{S}(T) \tag{5.45}$$

and

$$\lambda_{\phi}^{G}(T) = e^{-\phi(T)} \lambda_{\phi}^{S}(T) \,. \tag{5.46}$$

These conditions ensure that the problems are properly related.

The Hamiltonian function for the outer problem is again the same as for the unconstrained model, taking into account that the change in vitality is now given by (5.35). Effectively, the constraint on the change in vitality is a constraint on the range of possible trajectories of π to the subset $\pi \epsilon [\pi_0, 1]$. Note that the second control variable T marks start and end conditions for the two connected problems but does not enter the associated Hamiltonian functions. Further note that if one applies conditions (5.43), (5.45) and (5.46) in the solutions for ψ, λ_{ψ} and λ_{ϕ} it turns out that the solutions remain the same for both problems.

The Hamiltonian for the constrained model is

$$H = H^{G} + e^{-\phi(T)} H^{S} \tag{5.47}$$

where H^G is different from zero in $[0, T)$ and H^S is different from zero in $[T, \infty]$. Both H^G and H^S are essentially the same except for $min/max[f, 0]$ (see (5.35) and (5.36)) and the different initial values for vitality. To find the optimal value for T it is not necessary, however, to go through the procedure of maximizing H w.r.t. T. There is a simpler way: Basically, one has two optimization problems that can be solved separately and be spliced together for any given value of $\psi(T)$. Solving the two problems for any $\psi(T)$ the optimal $\psi(T)$ can subsequently be found by simple maximization.

The big advantage of introducing the second control variable T is that for the modified model an optimal solution for vitality has to be monotone not for one whole but for two distinct time intervals corresponding to two distinct yet related optimization problems. Between age zero and age T vitality can increase and between age T and infinity vitality may go down but must not increase. Obviously, all solutions to the modified model are subsets to solutions of the initial, unconstrained model. As such, optimal solutions for the initial model are optimal solutions for the modified model. But for ranges of parameters that have no optimal solution in the unconstrained model there might well be an optimal solution in the constrained model.

One runs into the paradoxical situation that adding restrictions opens up opportunities. Though the modified model is a constrained version of the initial model it extends the range of the parameter space for which optimal solutions exist, and in this way it also extends the range of possible qualitative trajectories. In particular, the constrained model has a solution for the case where both η parameter exceed one, namely full growth at $\pi = 1$ followed by full reproduction at $\pi = 0$ accompanied by a decline in vitality and thus increasing mortality, eventually followed by maintenance at $\pi = \pi_0$, when mortality plateaus. Thus, the constrained model can explain senescent as well as non-senescent life history strategies while the unconstrained model can only explain non-senescence.

5.3.1 Expected Solutions

In the following I list the expected solutions for all subsets of the parameter space for combinations of η_r and η_g, depending on whether those parameters are smaller, equal or larger than one:

- **$\eta_r < 1$ & $\eta_g < 1$:** If both η's are smaller than one the Hamiltonian is concave in π. An optimal solution exists and intermediate investment is expected. Parallel growth and reproduction followed by maintenance should be optimal. The optimal strategy for the second control should be to not throw the switch ($T = \infty$). I will call this strategy *Enhancement* (one may also call it *Negative Senescence* as in Vaupel et al. 2004).

- $\eta_r < 1$ & $\eta_g = 1$ **or** $\eta_r = 1$ & $\eta_g < 1$ **:** If one of the η's is equal to one and the other is smaller than one, then the Hamiltonian is concave in π. Therefore Enhancement should be optimal.
- $\eta_r = 1$ & $\eta_g = 1$**:** If both η's are equal to one the Hamiltonian is linear in π. As discussed in Chap. 4 the optimal solution involves a period of full growth corresponding to $\pi = 1$ followed by maintenance at π_0. I will call this strategy *Sustenance.*
- $\eta_r > 1$ & $\eta_g > 1$**:** If both η's are larger than one the Hamiltonian is convex in π. As discussed above, my modified model yields optimal solutions corresponding to full growth ($\pi = 1$) followed by full reproduction ($\pi = 0$) followed by maintenance ($\pi = \pi_0$). I call this kind of strategy *Senescence.*
- $\eta_r > 1$ & $\eta_g = 1$ **or** $\eta_r = 1$ & $\eta_g > 1$**:** If one of the η's is equal to one but the other is larger than one the Hamiltonian is convex in π and Senescence should be optimal.
- $\eta_r > 1$ & $\eta_g < 1$ **or** $\eta_r < 1$ & $\eta_g > 1$**:** If one η is larger than one and the other is smaller than one, the balance between η_r and η_g determines whether the Hamiltonian is concave or not. These exotic cases need further investigation.

Exotic Strategies

In the beginning of this section I derived the condition that determines whether the Hamiltonian function is concave or convex (see inequality (5.26)). Strikingly, the Hamiltonian function H can be concave or convex, if one η is larger and the other is smaller than one, depending on the value of π. To see this more clearly one can equate both sides of inequality (5.26) to find the point π_c that separates the two regions of π that are associated with concave vs convex H:

$$\pi_c = \frac{1 - \eta_g}{\eta_r - \eta_g}, \tag{5.48}$$

lower case c indicating "cut between convex and concave". With a minute of thought one can establish that $0 < \pi_c < 1$, i.e. the cut point truly lies within the interval $[0, 1]$.

Interestingly, depending on whether η_r or η_g is the parameter that exceeds one, it is the interval to the left or to the right side of π_c that corresponds to concave H. This should have important implications for the associated optimal strategies. It should matter whether it is growth or reproduction that faces increasing vs. decreasing returns to scale.

From inequality (5.26) one can show that the concave zone of H lies to the left of π_c when η_r exceeds one. An optimal π^*-strategy that does not run into my constraint has to obey

$$\pi^* < \pi_c \text{ if } \eta_r > 1 . \tag{5.49}$$

Conversely, the concave zone of H lies to the right of π_c if η_g exceeds one. An optimal π^*-strategy that does not run into my constraint has to obey

$$\pi^* > \pi_c \text{ if } \eta_g > 1 . \tag{5.50}$$

A helpful insight can be gained from (5.48): As η_r approaches one the cut point π_c moves towards one,

$$\lim_{\eta_r \to 1} \pi_c = 1 . \tag{5.51}$$

Equations (5.49) and (5.51) together imply that – as η_r approaches the limit of 1 from above – the whole interval will become concave. The strategy converges to Enhancement. This is intuitively right since one is back to the well behaved case of $\eta_r = 1 \,\&\, \eta_g < 1$. For values of $\eta_r < 1$, (5.50) and (5.51) imply convergence to the case of $\eta_r = 1 \,\&\, \eta_g > 1$, where the whole interval corresponds to convex H, so the strategy converges to Senescence.

Analogously, it holds that

$$\lim_{\eta_g \to 1} \pi_c = 0 . \tag{5.52}$$

If η_g approaches 1 from below, then the strategy converges to senescence, since $\eta_g < 1$ means $\pi < \pi_c$, so the region where H is concave disappears as $\pi_c \to 0$. If η_g approaches 1 from above, then the strategy converges to Enhancement, since $\eta_g > 1$ means $\pi > \pi_c$, so the region where H is concave covers the whole interval.

These dynamics can also be seen from

$$\frac{\partial \pi_c}{\partial \eta_r} = -\frac{1 - \eta_g}{(\eta_r - \eta_g)^2} \begin{cases} < 0 \text{ if } \eta_g < 1 \\ > 0 \text{ if } \eta_g > 1 \end{cases} \tag{5.53}$$

and

$$\frac{\partial \pi_c}{\partial \eta_g} = \frac{1 - \eta_g}{(\eta_r - \eta_g)^2} - \frac{1}{\eta_r - \eta_g} \begin{cases} < 0 \text{ if } \eta_g < 1 \\ > 0 \text{ if } \eta_g > 1 . \end{cases} \tag{5.54}$$

What can be deduced about the shape of possible strategies? It is hard to predict the expected optimal solutions. Presumably, some part of a strategy should correspond to extreme cases of $\pi = 1$, or $\pi = 0$ followed by maintenance $\pi = \pi_0$ and the other half of the strategy could be intermediate investment, i.e. either parallel growth and reproduction or parallel shrinkage and reproduction. In the latter case, slower than exponential senescence would be optimal.

For $\eta_g > 1$ the well-behaved area is right of π_c (i.e. above π_c), thus larger values of π up to one correspond to concave H. Therefore, a smooth transition from $\pi = 1$ to lower values is expected. This suggests parallel growth and reproduction. Hence, the "good" part of the strategy would be in the beginning of life. Therefore, the second part of life might correspond to the extreme case where the constraint of my constrained model comes into action.

For $\eta_r > 1$ the well-behaved area is left of π_c (i.e. below π_c), thus smaller values of π down to zero correspond to concave H. Therefore, a distinct jump of π from $\pi = 1$ to some lower value is expected. Since the case of $\eta_r > 1$ mirrors the one of $\eta_g > 1$, I would deduce that now the second part of the strategy is the "good" one. Thus, full growth followed by slower than exponential senescence could be an optimal strategy.

The discussion above suggests that there are several stages of life. For my models, life always starts with a period of development in which an organism grows and mortality falls but there is no fertility. Then there are four possibilities: First, the organism could maintain itself at a constant level of vitality, mortality and fertility – I call this "sustenance". Second, the organism could start to reproduce but continue to grow, with declining mortality – I call this "enhancement". Third, the organism could reproduce as much as possible, with mortality rising exponentially – this I call "senescence". Fourth, the organism could reduce the increase in mortality and decline in vitality by diverting some resources from fertility to repair – I call this "subsenescence". In my vitality model, the first stage is always development and the last stage is always maintenance. Thus, possible life history stages resulting from my model are development, enhancement, senescence, subsenescence and maintenance.[12]

[12] Note that maintenance and Sustenance both pertain to cases of constant mortality and fertility. I use Sustenance with capital letter to describe a life history *strategy*, namely that of development followed by maintenance, whereas maintenance with lower case letter simply refers to one *phase* of a life history. Analogously, I use Enhancement and Senescence with capital letters to describe life history strategies while the same terms in lower case letters refer to phases of a life history.

5.4 Numerical Results

While it is possible to write down integral equations that can explicitly be solved for the cases where $\eta_r > 1 \,\&\, \eta_g > 1$, finding solutions for the remaining η parameter space is difficult to do analytically. To solve for the age-trajectories of optimal investment, vitality, mortality and fertility, I implemented an algorithm based on Bellman's approach of dynamic programming (see App. A.3).

Applying this algorithm[13] I gain particular solutions for each part of the η parameter space and can check, whether the general shape of my numerical results fits the expected solutions based on my theoretical considerations in Sect. 5.3.1.

5.4.1 The Five Varieties of Life History Strategies

Five different types of optimal strategies can be found to result from this model. They are:

- *Sustenance* – development followed by maintenance,
- *Enhancement* – development followed by enhancement followed by maintenance
- *Senescence* – development followed by senescence followed by maintenance
- *Subsenescence* – development followed by subsenescence followed by maintenance
- *Ensenescence* – development followed by enhancement followed by senescence followed by maintenance

Note that this variety of strategies includes senescent as well as non-senescent life histories as predicted by my theoretical considerations above. The following figures exemplify particular optimal life history strategies for particular parameter values of η_r and η_g to show the scope of possible solutions and to emphasize the major importance of those two parameters.

Strategies Without Senescence

As expected, parameter values that are below one lead to Enhancement (see Figs. 5.3, 5.4 and 5.5). Note that for $\eta_r = 1 \,\&\, \eta_g = 0.5$ the onset of

[13] Due to numerical approximation errors caused by step length in vitality and investment, the algorithm sometimes leads to solutions that are pseudo maintenance, i.e. very slow senescence, where π is almost π_0. This pseudo maintenance converges to true maintenance when step length for vitality and investment are more and more reduced.

reproduction is abrupt, investment "falls down a cliff" whereas for $\eta_r = 0.5$ & $\eta_g = 0.5$ as well as $\eta_r = 0.5$ & $\eta_g = 1$ the onset of reproduction comes smoothly. Thus, if reproduction faces constant returns to scale in investment, reproduction starts abruptly while decreasing returns to scale imply a smooth transition (in agreement with my theoretical considerations in Sect 5.3.1).

If both parameters equal one, then - as predicted - Sustenance is optimal, as can be seen in Fig. 5.6.[14]

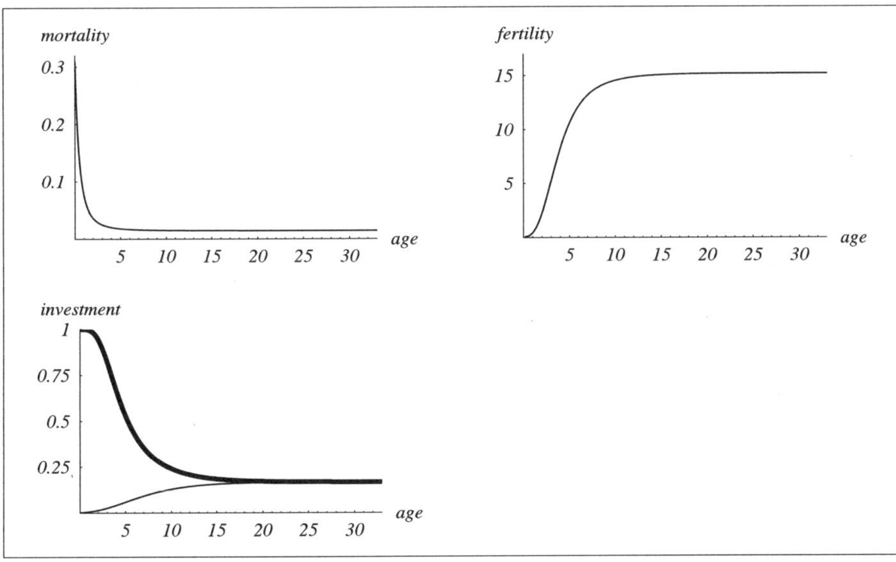

Fig. 5.3. Example Enhancement: $\eta_r = 0.5$ and $\eta_g = 0.5$

Strategies with Senescence

Strategies with senescence emerge as the η parameters start to exceed one as exemplified in the following figures. Note the difference between Fig. 5.7 and Fig. 5.8: a higher value of η_r leads to lower vita-

[14] For this and all the following figures in this section, the thick line in the lower left graph depicts the optimal investment strategy across age and the thin line depicts the corresponding level of investment π_0 required for maintenance.

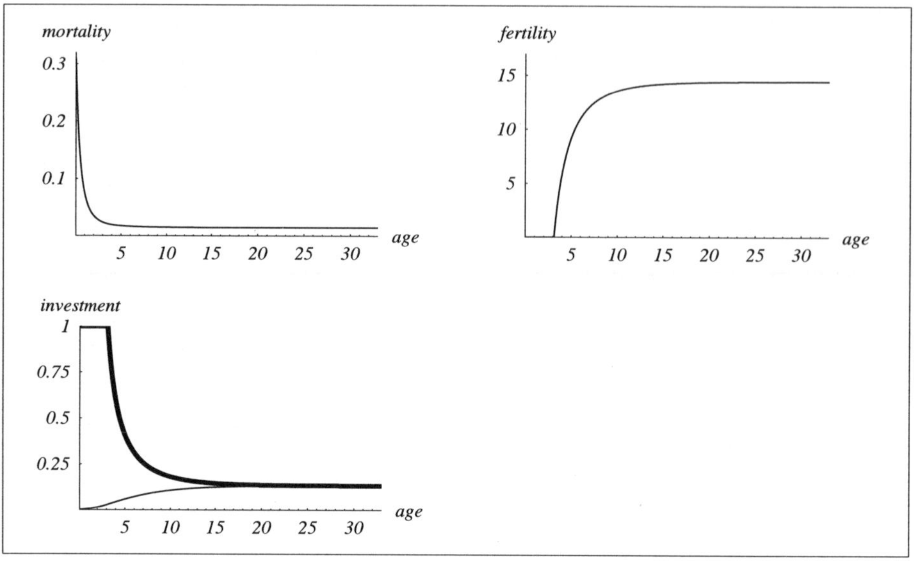

Fig. 5.4. Example Enhancement: $\eta_r = 1$ and $\eta_g = 0.5$

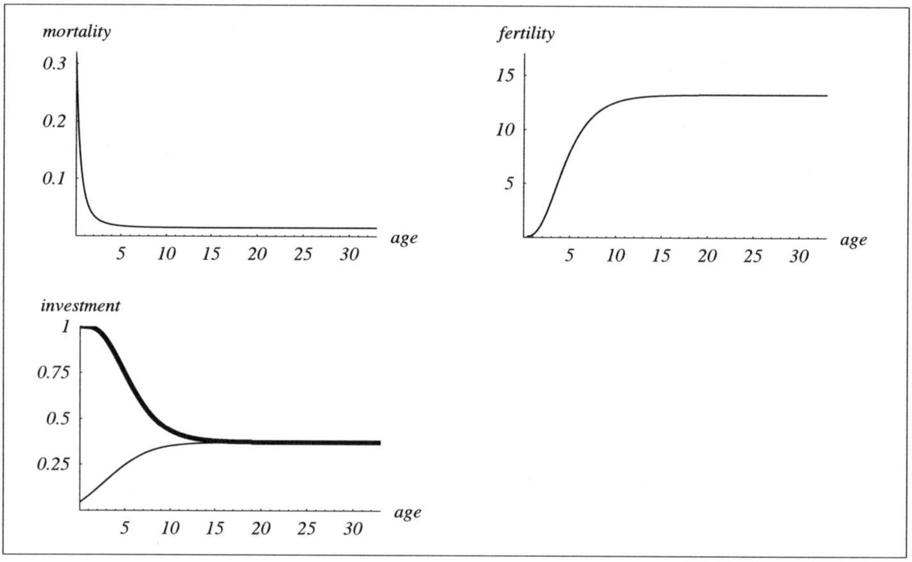

Fig. 5.5. Example Enhancement: $\eta_r = 0.5$ and $\eta_g = 1$

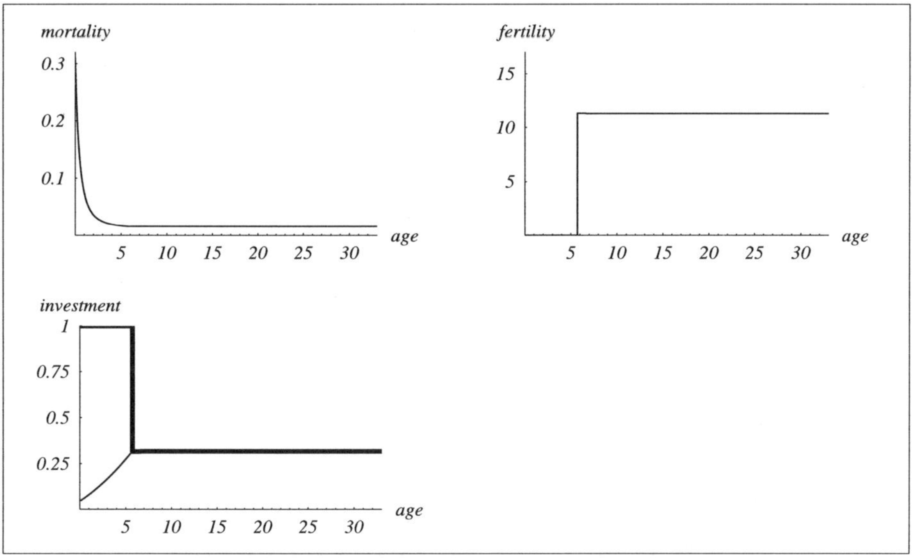

Fig. 5.6. Example Sustenance: $\eta_r = 1$ and $\eta_g = 1$

lity at maintenance and thus higher mortality. Interestingly, mortality increases only a little bit (relative to initial mortality[15] of 0.3) before reaching a plateau. For both η's equal to 3, Fig. 5.9 depicts the associated strategy (note that the time axis is double as long as in the previous figures). If the time units corresponded to years this setting of parameters captures a life history strategy of with humanesque features. Mortality falls until the age of maturity at about 13. Thereafter, mortality rises exponentially at a constant rate $\delta = 0.1$. Reproduction follows a hump-shaped curve. At the age of 60 reproduction drops to close to zero, corresponding to its level at the plateau. Note that virtually all lifetime reproduction is realized before this age and further contributions of later age classes are negligible. My model does not lead to true menopause – reproduction continues albeit at a very low level, and mortality remains constant. In a human life history reproduction, at least for females, should cease and mortality should keep on rising exponentially. Clearly, a model as simple as mine that is solely based on vitality can only capture a rough, humanesque pattern.

[15] Note that in Figs. 5.7, 5.8, 5.10, 5.11 and 5.13 the increase in mortality is barely visible because mortality is shown on a scale set by initial mortality, which is at the magnitude of 0.3. It should be emphasized, however, that the force of mortality is increasing substantially in all these figures relative to its level at reproductive maturity.

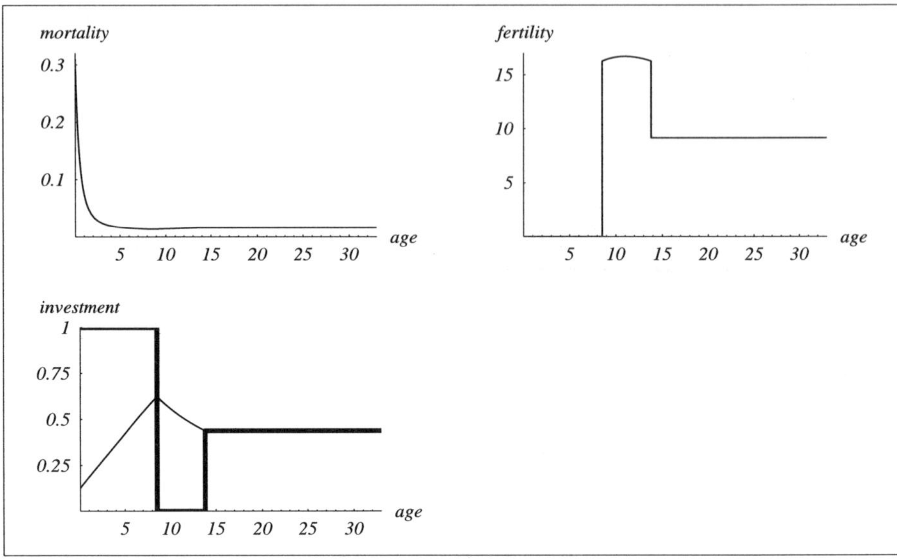

Fig. 5.7. Example Senescence: $\eta_r = 1$ $\eta_g = 1.5$

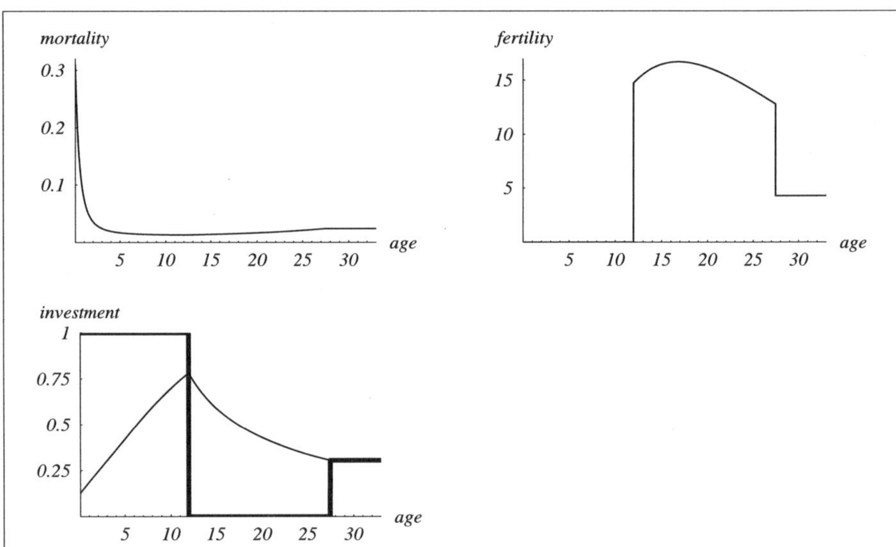

Fig. 5.8. Example Senescence: $\eta_r = 3$ $\eta_g = 1.5$

Ensenescence

Figures 5.10, 5.11, 5.12 and 5.13 show results that exemplify parameter combinations corresponding to areas where exotic strategies were

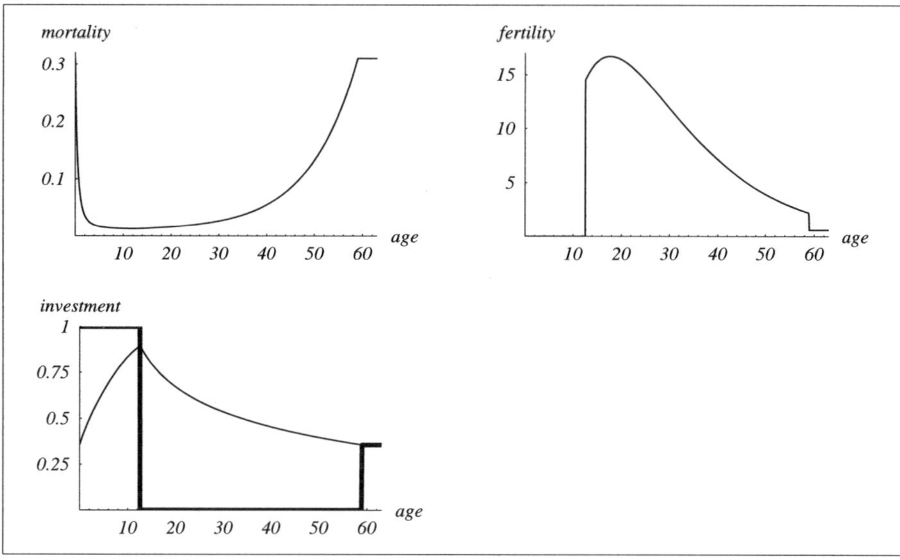

Fig. 5.9. Example Senescence (humanesque case): $\eta_r = 3$ $\eta_g = 3$

expected. And indeed, the strategies are really interesting: The combination of $\eta_r = 0.5$ & $\eta_g = 1.5$ and $\eta_r = 0.5$ & $\eta_g = 3$ (Figs. 5.10 and 5.11) depict strategies that include parallel reproduction and growth. An initial period of development ($\pi = 1$) is followed by a period of parallel growth and reproduction ($\pi > \pi_0$), leading into a period of exponential senescence due to a drop in the strategy to $\pi = 0$ and eventually reaching a level of vitality that will be maintained ever after ($\pi = \pi_0$). I call this strategy *Ensenescence* because it includes a period of enhancement and a period of senescence. Figure 5.11 shows the effect of increasing η_g, which is to widen the period of senescence and flatten the dip.

Note that the dip in the strategy depicted in Figs. 5.10 and 5.11 is not an artefact. I calculated many different η combinations for the case of $\eta_r < 1$ and $\eta_g > 1$ and always found this peculiar strategy.

Senescence and Non-senescence Close Together

The range of parameters $\eta_r > 1$ and $\eta_g < 1$ reveals a surprise: For $\eta_r = 1.5$ & $\eta_g = 0.5$ (Fig. 5.12) the optimal strategy is Enhancement. Investment shows a cliff at the onset of reproduction as predicted from $\eta_r > 1$ and the discussion of π_c. This kind of enhancement differs from the very early and gradual start of reproduction shown in Figs. 5.3–5.5

Fascinatingly, there is yet another strategy to be found in this parameter region. Increasing η_r to 3 as shown in Fig. 5.13 moves the non-senescent life history towards a senescent one. Parallel reproduction and growth is not favored anymore. Instead the optimal strategy is development ($\pi = 1$) followed by slower than exponential senescence[16] ($0 < \pi < \pi_0$), followed by maintenance ($\pi = \pi_0$). I call this strategy *Subsenescence.*

If the optimal strategy is Enhancement the vitality trajectory is monotone increasing (and my model constraint is not in action). If, however, Subsenescence is optimal the vitality trajectory first increases and then decreases (my model constraint applies)[17]. Thus, for increasing returns to scale in reproduction but decreasing returns to growth and maintenance, both senescent and non-senescent strategies can be optimal.

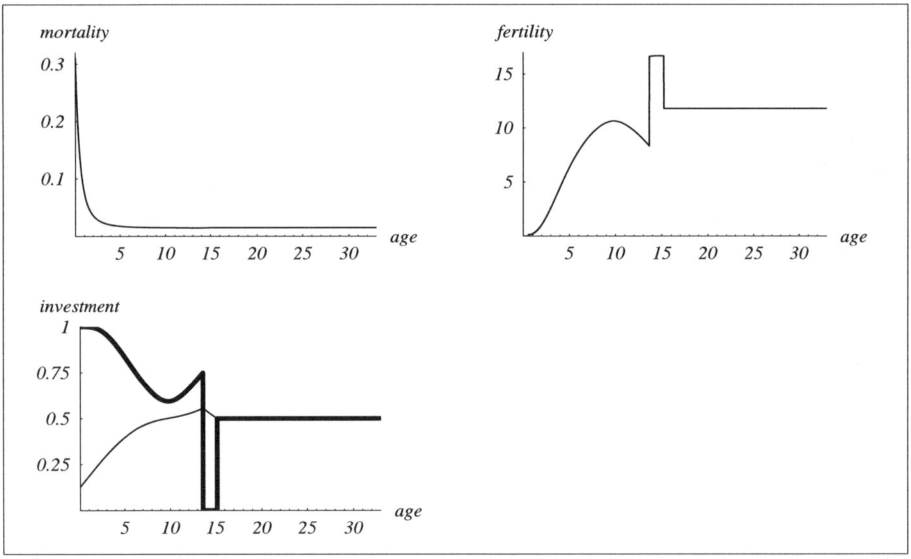

Fig. 5.10. Example Ensenescence: $\eta_r = 0.5$ $\eta_g = 1.5$

[16] This is a true strategy and not an artifact, the strategy is robust with decreasing step length.

[17] The cliff when π drops from one to below π_0 shows that the investment trajectory jumps over the convex area of π's which lies above π_c for this range of η parameters.

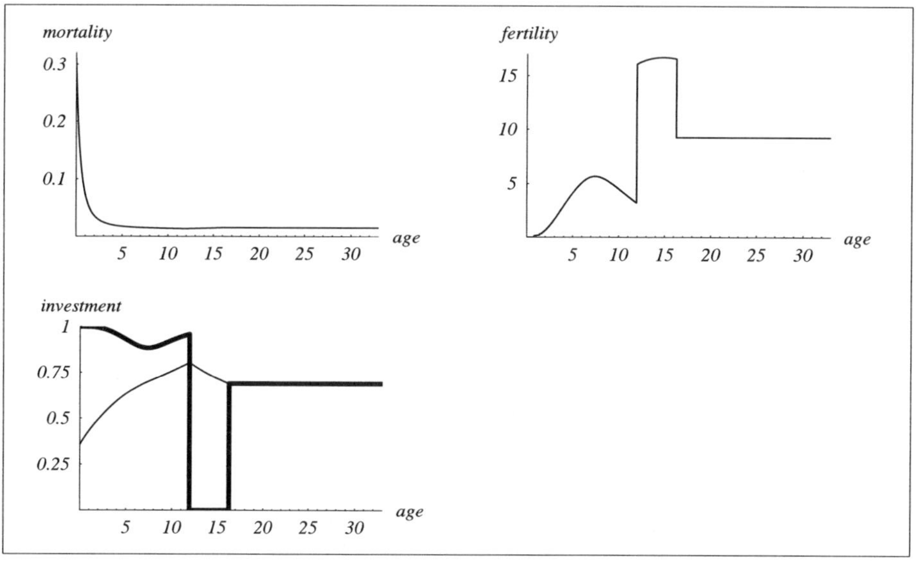

Fig. 5.11. Example Ensenescence: $\eta_r = 0.5$ $\eta_g = 3$

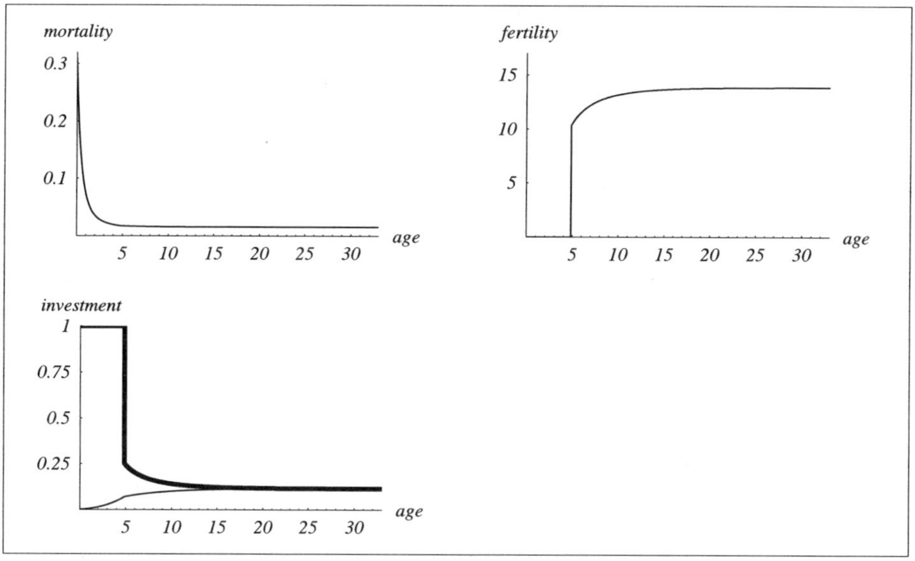

Fig. 5.12. Example Enhancement: $\eta_r = 1.5$ $\eta_g = 0.5$

Strategies Across η Parameters

The matrix of strategies in Fig. 5.14 shows how strategies change across the range of η_r and η_g. Each graph displays the optimal investment

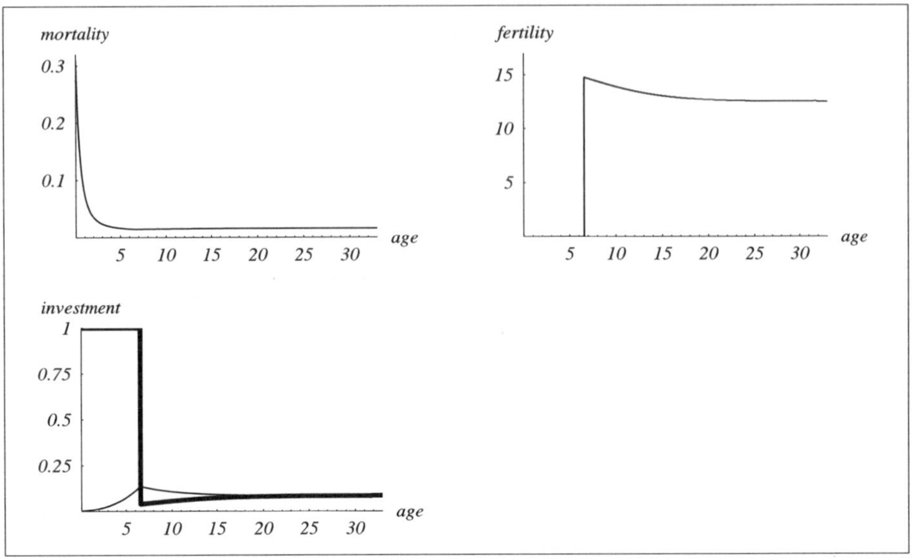

Fig. 5.13. Example Subsenescence: $\eta_r = 3$ $\eta_g = 0.5$

strategy π^* across age. In each graph, initially π^* equals one and drops below one at the age of maturity[18].

Consider for instance the first column, with η_g fixed at 0.5. As η_r increases from 0.5 to 1.5, the onset of reproduction becomes more abrupt. When η_r is 3.0 the optimal strategy is Subsenescence rather than Enhancement: there is no longer a period of parallel reproduction and growth.

Another interesting example concerns the upper right corner of Fig. 5.14. The strategy in the corner is Ensenescence. The name "Ensenescence" combines two strategies – Enhancement and Senescence, and it is those two strategies that Ensenescence converges to as either of the η parameters changes. My computational results (not presented here) show how a reduction in η_g from 1.5 to 1 drives the strategy towards Enhancement. As η_g decreases the valley of the dip becomes more and more pronounced, eventually flattening out and the period of senescence becomes shorter and eventually disappears. When instead η_r increases from 0.5 to 1, Ensenescence converges to Senescence. The

[18] Note that the thin line that depicts the required investment for maintenance originating from the lower left corner of each graph starts off at higher and higher values as η_g increases along each row. The reason are higher costs of maintenance due to larger values of η_g (compare (5.7)).

dip becomes more and more shallow, eventually disappearing and the period of senescence lengthens[19].

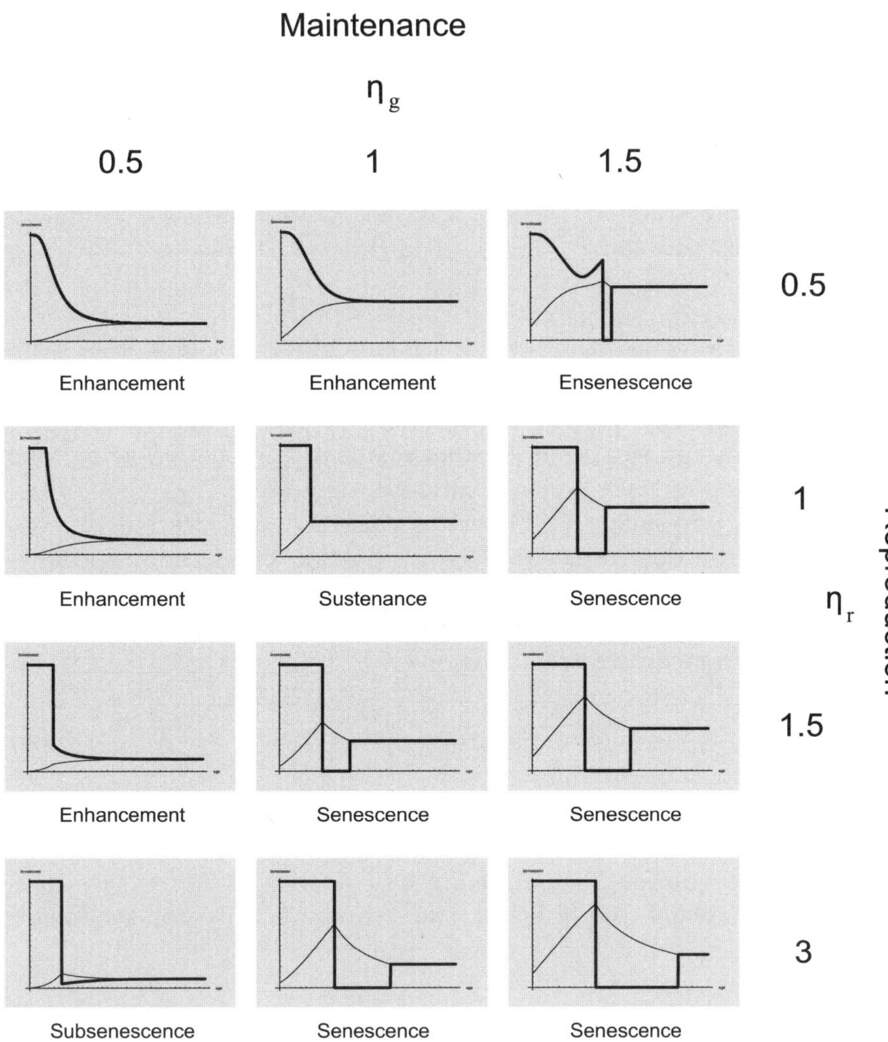

Fig. 5.14. Overview of optimal investment strategies across η parameters

---

[19] I believe that the increase in π just before the onset of senescence is due to the fact that the organism invests more in growth and survival to ensure survival to the beneficial phase of full reproduction.

5.4.2 When Senescence Is Optimal and when It Is Not

The previous section concentrated on qualitative age-patterns of mortality. Several of the mortality and fertility trajectories increase over some part of reproductive life but decrease or remain constant over other parts of reproductive life. How should one decide whether those life histories are senescent or non-senescent? How could one describe the 'degree of senescence' of a life history strategy? This section offers a measure of the degree of senescence and subsequently sheds light on the characteristics that determine whether senescent or non-senescent life histories are optimal.

The Degree of Senescence

My suggestion for a definition of 'senescence' in Chap. 1 applies to particular age-groups. What criteria should be used to label a complete life-history strategy senescent or non-senescent? In the following, one way of approaching such a classification is suggested.

Whether a particular life history is classified as senescent or non-senescent can be determined by the proportion of lifetime reproduction that is realized at ages when mortality rises, i.e. when $\pi < \pi_0$. This senescence indicator, S, measures the degree of senescence for a particular life history strategy. S is given by

$$S = \frac{\sum_{x=0}^{\infty} J_x \, l_x \, m_x}{\sum_{x=0}^{\infty} l_x \, m_x} \,, \tag{5.55}$$

where $J_x = 1$ if investment in growth is below maintenance level ($\pi(\psi(x)) < \pi_0(\psi(x))$) and $J_x = 0$ if investment is greater than or equal to the amount required for maintenance of vitality. If $S = 1$, the strategy is fully senescent and if $S = 0$, the strategy is fully non-senescent. All values in between describe mixed strategies.

The Crucial Parameters

The crucial parameters that are responsible for the qualitative shape of an optimal life history strategy are the η parameters and the mortality parameters.

Figure 5.15 shows how different levels of mortality can influence the degree of senescence in a particular life history strategy depending on the range of η parameters. Each combination of η's corresponds to one graph. Each graph displays the degree of senescence, i.e. values of S

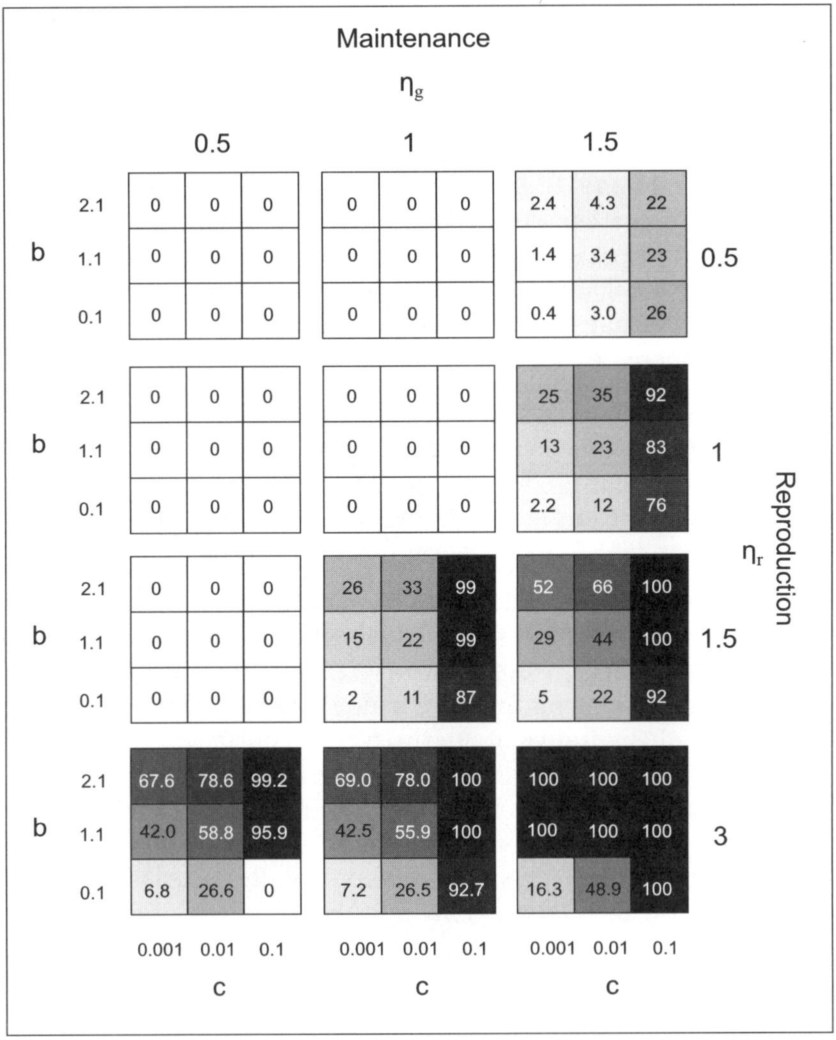

Fig. 5.15. Degree of senescence S, in%, across η and mortality parameters; *Color code: darker = more senescent; numbers refer to percent life time reproduction realized at ages when mortality increases*

across extrinsic[20] mortality c along the x-axis and intrinsic mortality b along the y-axis. Two general results can be derived from this figure:

- The upper left triangle of Fig. 5.15 is white; those ranges of parameters are associated with pure non-senescence, i.e. $S = 0$. Towards the lower right corner the areas become more and more shady and dark. Thus, higher levels of the η parameters generally correspond to higher degrees of senescence.[21]
- Inside each of the 12 graphs, lighter areas can be found towards the left and further down. Hence, higher levels of mortality – be it extrinsic (on the x-axis) or intrinsic (on the y-axis) – go hand in hand with higher degrees of senescence, in other words a larger fraction of lifetime reproductive success is realized during ages when mortality increases.

The influence of the η and mortality parameters on age and vitality at maturity can be seen in Figure 5.16, which follows the same logic as the previous figure. Clear gradients in shading from light to dark become apparent: from left to right, top to bottom and – inside each graph – right to left. Three general observations can be made:

- Higher levels of overall mortality are generally associated with smaller vitality at maturity, higher extrinsic mortality c in particular has this effect. Note, however, that higher intrinsic mortality b affects vitality at maturity in a nonlinear fashion: intermediate b leads to higher vitality at maturity, i.e. continued growth pays off via better survival at maturity. Too high a level of b, however, counterbalances this advantage because of increasing overall mortality – earlier maturation at a smaller level of vitality is favored.
- Higher levels of the η parameters are generally associated with larger vitality and thus a later age at maturity.
- For non-senescent strategies (white areas), vitality at maturity barely exceeding 50 % of the maximum vitality possible, i.e. development does not proceed beyond the point where the increase in vitality starts to slow down (The age-trajectory of vitality is s-shaped for $\pi = 1$). For senescent strategies vitality at maturity is generally higher than for non-senescent strategies, reaching about 75% and more percent of maximum vitality. Thus there is a distinct shift in vitality at maturity relative to maximum attainable vitality

[20] In this monograph I use the term "extrinsic" meaning "state-independent" and "intrinsic" meaning "state-dependent".

[21] A forth column for $\eta_g = 3$ would reveal the same pattern, mainly being dark; it is not displayed here.

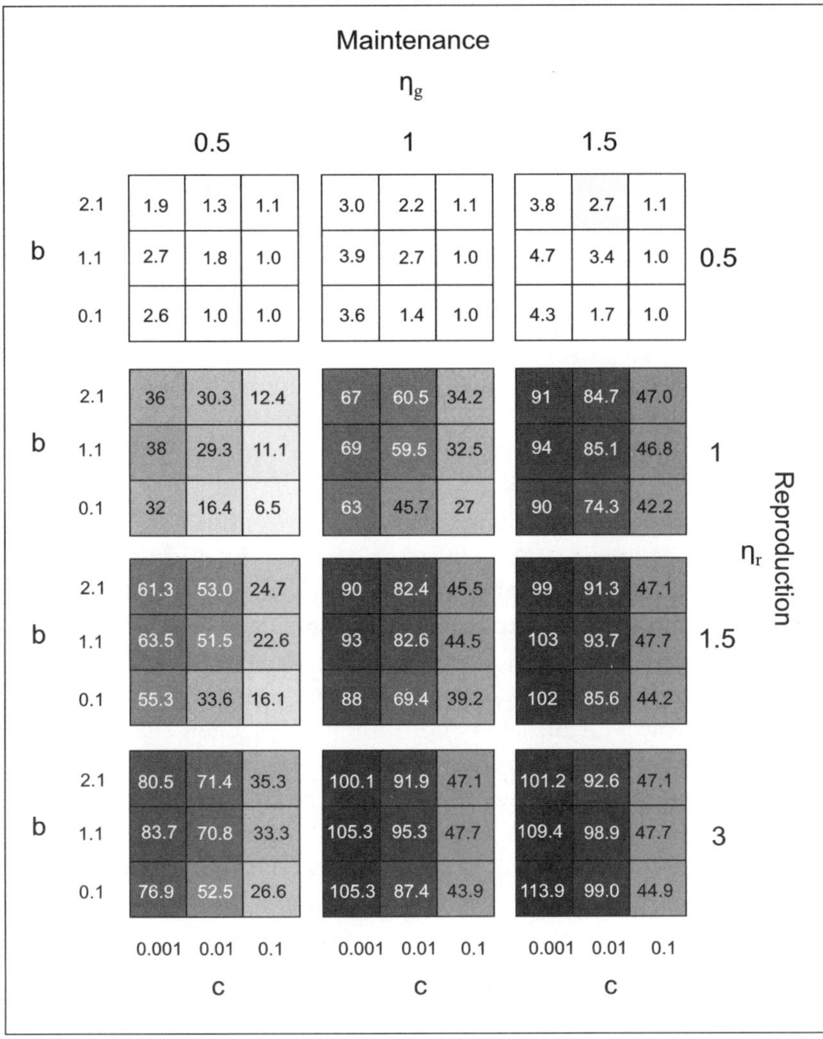

Fig. 5.16. Vitality at maturity across η and mortality parameters; *Color code: darker = larger vitality at maturity*

Ψ between senescent and non-senescent strategies(as can be seen from the distinct shift between light and dark areas).

In short, higher levels of the η parameters correspond to a higher degree of senescence as well as greater vitality at maturity. Higher levels of mortality lead to higher degrees of senescence but reduce vitality at maturity. Non-senescent strategies are associated with relatively[22] low vitality and thus early age at maturity, while senescent strategies are associated with relatively large vitality and thus late age at maturity.

A Peculiar Little White Square

Figure 5.15 reveals a peculiarity: the lower right little square of the lowest left graph is white, there is no senescence. Given the patterns elsewhere in the figure one would expect a value of S above 80. While all strategies for $\eta_r = 3$ and $\eta_g = 0.5$ are Subsenescence, the highest value of $c = 0.1$ at the lowest value of $b = 0.1$ corresponds to Enhancement. Why?

First note that all except one (namely Ensenescence) of my strategies exhibit monotone state trajectories after reproductive maturity. Thus, for those strategies the relative level of vitality at maturity to vitality at maintenance can predict whether vitality increases or decreases across adult ages. If vitality at maturity is larger than vitality at maintenance, the organism must have been shrinking after maturity to reach that lower level, thus there must have been senescence. If vitality at maturity is smaller than vitality at maintenance, the organism must have been growing after maturity to reach that higher level, thus there must have been non-senescence.

If one secondly recalls that non-senescence implies small and senescence implies large vitality at maturity, then it seems that vitality at maturity and vitality at maintenance hold the key to explain the peculiar little white square.

Influence of η Parameters

The influence of the η parameters on vitality at maturity and at maintenance is shown in more detail in Figs. 5.17 and 5.18 [23]:

[22] "Relative" here means relative to maximum vitality.

[23] Both figures are calculated keeping all other parameters unchanged at $b = 0.3$, $c = 0.01$, $k = 3$, $\kappa = 0.8$, $\delta = 0.1$ which correspond to a maximum attainable vitality of 123. Remember that vitality is a dimensionless variable so the specific value of 123 only means that maximum vitality is 123 times greater than vitality at age zero.

- Optimal vitality at maturity increases as either η_r or η_g increases, i.e. the steeper the returns to scale in either reproduction or growth and maintenance (stronger convexity), the later the age at maturity and the higher the acquired reproductive potential. Figure 5.17 visualizes this finding, both by reading the figure across the four lines from thinnest to thickest and by looking at each line from left to right.
- The ultimate level of vitality that will be maintained decreases as either η increases, i.e. the more convex the returns to scale in either reproduction or growth and maintenance, the later the age at maintenance and the lower the reproductive potential ultimately maintained. Figure 5.18 visualizes this finding, again both by reading the figure across the four lines from thinnest to thickest and by looking at each line from left to right.

Thus, for low values of the η's maturity happens early at a low level of vitality while a much higher level of vitality eventually is maintained. For high values of the η's, maturity is postponed, hence vitality at maturity is large, but the level that is eventually maintained is small. Consequently, there is a region of η parameters where vitality at maturity is smaller than vitality at maintenance, and there is a region of η parameters where vitality at maturity is larger than vitality at maintenance. This implies that there is a cut point where vitality at maturity equals vitality at maintenance. Figure 5.19 makes this verbal argument clearer by displaying the results from Fig. 5.17 and 5.18 together. The cut points for lines of equal thickness correspond to the values of η_r and η_g that lead to strategies where vitality at maturity equals vitality at maintenance[24].

The influence of the mortality parameter c on vitality at maturity and vitality at maintenance is shown Figure 5.20. Clearly, both vitality at maturity and vitality at maintenance decline as extrinsic mortality c increases, but since vitality at maturity declines faster than vitality at maintenance, there is a level of c where maturity and maintenance cross. At this cut point, Sustenance is the optimal strategy. For values of c smaller than the cut point, vitality at maturity exceeds vitality at maintenance, which implies a falling trajectory of vitality. Subsenescence is optimal. For values of c larger than the cut point, vitality at maturity falls below vitality at maintenance, which implies an increasing trajectory of vitality. Enhancement is optimal. Note that the curves

[24] Analogous figures not displayed here can be calculated for the case when η_g runs across the x-axis while different levels of η_r are given by lines of different thickness.

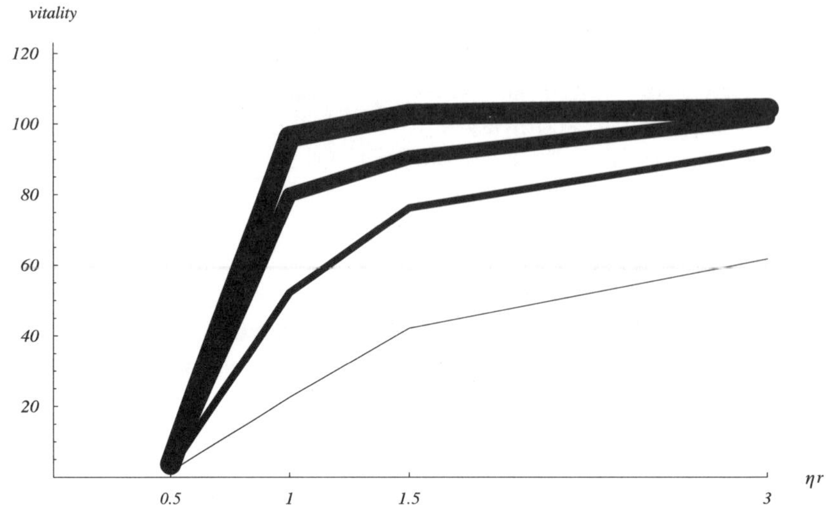

Fig. 5.17. Vitality at maturity across four distinct values of η_r plotted for the same four values of η_g; *line thickness being proportional to the value of* η_g

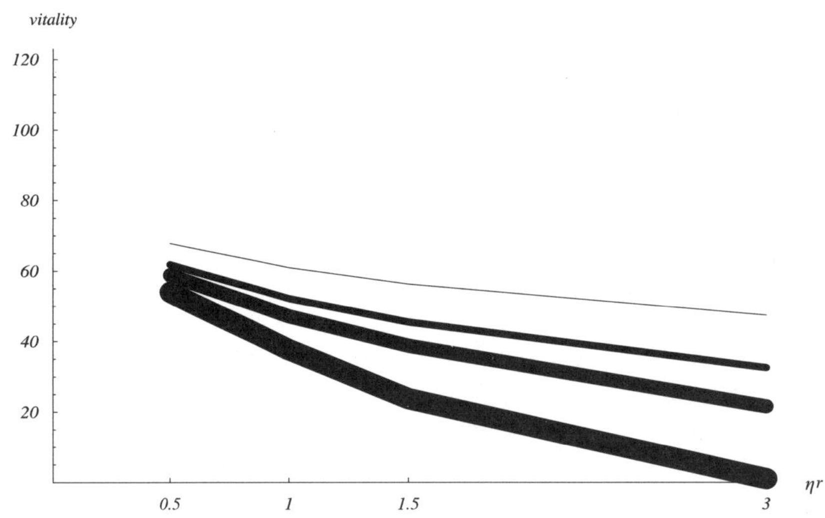

Fig. 5.18. Vitality at maintenance across four distinct values of η_r plotted for the same four values of η_g; *line thickness being proportional to the value of* η_g

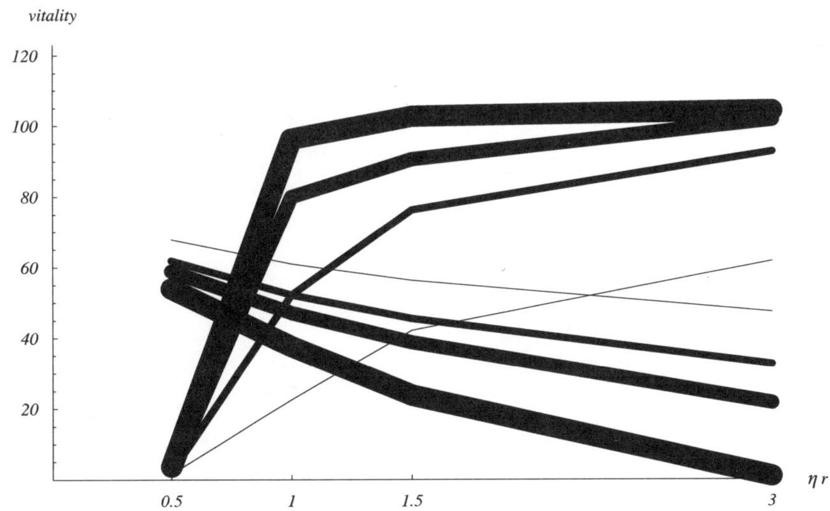

Fig. 5.19. Vitality at maturity and vitality at maintenance across four distinct values of η_r plotted for the same four values of η_g; *line thickness being proportional to the value of* η_g

for different values of b converge for large values of c, as the level of c dominates total mortality, i.e. when b is small compared to c [25].

In sum, the peculiar white little square is not so peculiar anymore. The explanation is straightforward: c can shift a strategy from Subsenescence to Enhancement by shifting vitality at maturity from above to below vitality at maintenance. At the cut point, where vitality at maturity is also the level of vitality that is maintained, Sustenance is optimal. Thus, a change in the extrinsic mortality parameter c can change a strategy from senescent to non-senescent, at least for the part of the η parameter space where $\eta_r > 1$ & $\eta_g < 1$.[26] The transition at the boundary between senescence and non-senescence is smooth, when caused by a change in c, contrary to the transition caused by changes in the η parameters. At the boundary between senescence and non-senescence that is set by the η parameters, the shift in vitality at maturity is distinct (compare Fig. 5.16).

[25] For values of c larger than those depicted in Fig. 5.15, non-senescence would have emerged in the graphs for values of $\eta_r > 1$ & $\eta_g < 1$. It is only because the range of c is not wide enough that this pattern did not become more apparent in this picture.

[26] From the analytic section in the beginning of this chapter it is clear that this range is the only one where mortality could shift the strategy from senescent to non-senescent

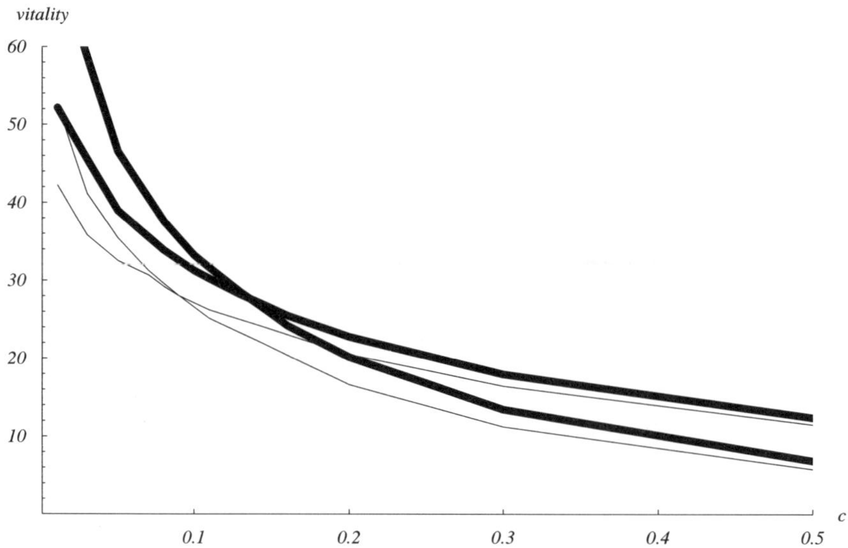

Fig. 5.20. Vitality at maturity and vitality at maintenance are plotted across extrinsic mortality c for values of parameter $b = 0.1$ and $b = 1.1$. Since the curves for $b = 2.1$ turned out to be very similar to those for $b = 1.1$, the case $b = 2.1$ is not shown. The pair of thick lines correspond to $b = 1.1$, the pair of thin lines correspond to $b = 0.1$. The initially upper curve of each pair depicts vitality at maturity, and the initially lower curve of each pair depicts vitality at maintenance.

5.5 Discussion

Whether a strategy follows a senescent or a non-senescent path is crucially dependent on the η parameters.

5.5.1 Senescence vs. Non-senescence

Non-senescent strategies are favored when the η parameters are below or equal to one. Values below one imply that returns to investment are decreasing, i.e. concave. An alternative verbal interpretation of decreasing returns to scale would be "cheap" or "efficient" investment, since a substantial amount of output can be realized with a small fraction of input. Thus, organisms with growth, maintenance and reproductive systems that can be used efficiently with only a fraction of total resources should follow non-senescent life history strategies.

Senescent strategies are favored when none of the two η parameters is below one and at least one of the two η parameters is above one. Values above one imply that returns to investment are increasing, i.e. convex. An alternative verbal interpretation of increasing returns to investment would be "costly" or "inefficient" investment, since a significant amount of output can only be realized if all resources are concentrated on one process exclusively. Thus, organisms with growth, maintenance and reproductive systems that work best if used successively should tend to adopt senescent life history strategies.

Mixed strategies that include aspects of both senescent and non-senescent life histories can be found if one η- parameter is above and the other below one, i.e. when returns to investment are decreasing (concave, cheap, efficient) in one and increasing (convex, costly, inefficient) in the other process - either reproduction or growth and maintenance.

If growth and maintenance are efficient, then strategies can be either non-senescent (mortality never increases) or partially senescent (mortality increases slower than exponentially for some time after reproductive maturity) depending on the extrinsic hazard of death.

If reproduction is efficient, then it is optimal to grow and reproduce simultaneously for some time after reproductive maturity. Following this non-senescent phase of life, it is favorable to concentrate all resources on reproduction and let mortality increase exponentially (eventually levelling).

5.5.2 The η Parameters in Nature

How could one identify a species' η- parameter range? Species that show concave returns to scale in both reproduction and growth are species that can easily share resources between those processes and that do not gain much by specializing in either one of them. Organisms that are capable of vegetative propagation, where growth can be considered an investment in reproduction are candidates for this category. It is important to note, however, that species with the ability for this asexual mode of reproduction also have the ability to reproduce sexually. I believe that the values of η_r, i.e. the returns to investment in reproduction, are significantly different from each other for asexual vs. sexual reproductive mode. If asexual reproduction is associated with concave investment and sexual reproduction with convex investment, then the former case is associated with non-senescence and the latter case with senescence. An example is *Hydra oligactis* that sustains its state unless it starts to reproduce sexually [220]. Understanding the different returns to scale for asexual and sexual reproduction and its implications for senescence vs. non-senescence of a species is one interesting avenue for future research.

Non-senescence is part of a species's life history if at least one of the η parameters is below one. Life histories with at least one η parameter below one should involve a period of simultaneous growth and reproduction. All indeterminately growing species, as for example most trees and many fish fall into this category. Whether those species show exponential senescence later in life or not depends on whether growth and maintenance or reproduction is efficient.

Convexity or concavity in reproduction can be distinguished as follows: if organisms face convexity in reproduction, i.e. $\eta_r > 1$, then the onset of reproduction should be sudden and pronounced. If on the other hand $\eta_r < 1$, then the transition into adult ages happens smoothly. An analogous prediction for growth and maintenance can not be derived from my model.

An indication for concave (cheap and efficient) reproduction might also be given by number and size of offspring. Many small offspring (each single offspring only contributing an iota to life time reproduction) could indicate concave reproduction and few large offspring (each offspring requiring 'heavy' investment and contributing substantially to life time reproduction) could indicate convex reproduction.

Convexity with respect to reproduction could also mean that it is initially costly to build the required machinery that is necessary to reproduce at all. Lavish reproductive structures like huge fancy flowers, long stalks in bamboo, or feeding structures inside and outside the mothers body in animals could be an indication.

Convexity in growth and maintenance could mean that the machinery for growth and maintenance can only keep going efficiently if all energy is used for that purpose.

An efficient, cheap way of maintenance is "throw away and grow new". Disposing of damaged tissue does not require a lot of energy, indeed it should not cost much at all. If lost tissue can easily be replaced at reasonable costs without disturbing an organism's functioning and integrity it may be a cheaper strategy than repairing the damage that occurred. Repair of existing structure is costly because it requires error recognition, knowledge of the undamaged state that is about to be restored and the appropriate machinery to do so. Loosing body parts or tissue (like leaves in a plant) might not need much information or energy at all but happen more or less automatically. Replacement of parts and tissue (in case of the leaves) might only need the template for making that part or tissue but can do without any knowledge about the damaged one. The signal for producing a new leaf, for instance, might solely require the signal of too little overall energy production by the remaining leaves.

Therefore organisms that are made of simple, repeated structure that can easily be discarded and regenerated when damaged, are candidates for concave returns to investment in maintenance and growth.[27] A thorough understanding of the $\eta-$ parameters in nature is an important avenue for future research.

5.5.3 The Mortality Paradox

In this section I further consider the case when reproduction is costly and growth and maintenance are cheap ($\eta_r > 1$ and $\eta_g < 1$). I found that an increase in parameter c, which captures state-independent mortality, can shift a strategy from Subsenescence to Enhancement. It appears that non-senescence is favored more strongly, the greater the extrinsic hazard. This is striking. Exactly the opposite has generally been stated – that a high risk of extrinsic death should favor senescence [212]. But my model predicts that this hypothesis is not true for species with low costs of maintenance and high costs of reproduction. What could explain this unexpected and seemingly paradoxical result? I addressed this above in terms of vitality at maturity vs. at maintenance. Here I take a different perspective.

It is a well-supported [169, 186] and intuitively appealing fact that a high extrinsic hazard of death favors early reproductive maturity. A short juvenile period reduces the time available for development and hence the time to attain a certain vitality. Vitality, however, determines the level of energy available and therefore the potential to reproduce. If individuals have to mature early because of a very risky

[27] Another example for repeated, easily replaced parts are red blood cells. Each cell wears out over its average lifetime of 120 days in humans, but new undamaged red blood cells are constantly provided by the bone marrow. Thus, there is senescence at the level of the red blood cells, but there could be sustenance at the next higher level of organization – the blood – if the age- and damage-structure in the population of red blood cells is constant, at least over a long period of a human's life. But at the whole organism level, the human being senesces. This further leads to the idea of looking at senescence vs. non-senescence at different levels of organization of an organism. Different levels may have different η parameters. The individual red blood cells might face increasing returns to scale, but the blood itself may have decreasing returns to scale while the whole body is subject to convex investment. Also, the value of the η parameters might change with age, being concave early in live and transitioning to convex later in live. For now this remains speculation and is far beyond the scope of this chapter. But it points to exciting research questions for future research.

environment, their reproductive potential might be small. The short life to be expected gives only few opportunities to reproduce. Thus, every additional reproductive event increases total lifetime reproduction by a large relative share compared to what has been realized before. Therefore, depending on the costs of reproduction, a small potential should be maintained (Sustenance) and, if possible, further increased (Enhancement) when maintenance costs are low ($\eta_g < 1$).

If, on the other hand, life is safe the individual can afford to spend a long time building up a high level of vitality, i.e. a large reproductive potential. Instead of paying the price of maintaining a high level of vitality, it may be evolutionarily advantageous to harvest this potential at the cost of a loss in functioning. Subsenescence is the strategy that particularly suits this circumstances. It is Subsenescence and not Senescence that is the optimal strategy at low levels of extrinsic mortality when maintenance is cheap but reproduction is expensive. And this is why: The propensity to share resources between reproduction and growth is small due to costly reproduction. Therefore exclusive investment is desirable. Low maintenance costs, on the other hand, favor the preservation of vitality rather than decay, which implies sharing of resources. As long as mortality is low the individual can afford to mature late, attaining a high reproductive potential. However, maintaining this level of vitality would be strongly penalized in terms of reduced reproduction. Instead, the individual harvests the large potential and mortality increases after reproductive maturity. But when vitality has fallen to a level that can be preserved without too much penalty, any further deterioration is suboptimal. The individual maintains its state and mortality is constant.

Williams [212] conjectures that low levels of extrinsic mortality should be associated with slow-senescent strategies and high levels of extrinsic mortality should be associated with fast-senescent strategies. His hypothesis is in accordance with the results from previous reproductive effort models (for a review see Charlesworth [27, Section 5.3.4.]). Higher extrinsic risk tends to increase reproductive effort, which implies higher levels of mortality. I have shown that my results predict under some circumstances the opposite effect of an increase in parameter c. Moreover, my results imply that non-senescent strategies can be optimal. A theory based on optimization of trade-offs can account for constant or declining age-patterns of mortality while a theory based on mutation accumulation cannot explain these patterns. In Sects. 6.3.1, 6.4.1, and 6.4.3, I will discuss the concept of extrinsic mortality and return to Williams's hypothesis.

5.5.4 Plateaus

All my life history strategies show maintenance from a certain vitality onwards, i.e. plateaus in mortality and fertility are part of all life history strategies. Plateaus have been observed in large populations of Medflies, *Drosophila*, nematode worms, beetles, and humans ([21], [47], [189], [201]), and understanding the cause of this pattern has caught the attention of many researchers (for example [32], [151], [201], [208], [209]).

In my model plateaus naturally arise as part of non-senescent life history strategies, since vitality cannot increase indefinitely and thus at some level will have to be maintained. The plateau that follows a period of exponentially increasing mortality results from the constraint I impose that vitality can not follow a zig-zag path. Therefore, even though my model can lead to mortality plateaus one has to be careful in interpreting this technical result. From a biological perspective there may be organisms that cannot halt further deterioration after they have deteriorated substantially. In such species, individuals may continue to suffer senescence with age; observed plateaus maybe due to population heterogeneity [201, 202, 204, 206, 207]. On the other hand, it may indeed be the case that mortality plateaus for some species are due to a strategy of sustenance at older ages. Research is needed.

5.6 Summary

The simple model developed in this chapter captures the main features of life: mortality, reproduction, development, growth and maintenance. The results show that the range of optimal life histories is wide. Senescent as well as non-senescent life history strategies can be optimal.

Whether an optimal life history follows a non-senescent strategy or a senescent strategy is crucially determined by the returns to scale to growth and maintenance as well as to reproduction. Efficient maintenance and growth systems favor maintenance strategies after growth is completed while efficient reproductive systems favor strategies of parallel growth and reproduction.

Senescent and non-senescent strategies show distinct differences in the level of vitality and age at maturity: Non-senescence is associated with early maturity at relatively low vitality and senescent strategies are associated with late maturity at a level of vitality that is close to its maximum attainable level. Since vitality at maturity can reasonably be interpreted as size at maturity, my vitality-based model leads to a hypothesis that also followed from my size-based model: species that mature early at a relatively small size with the potential of growth and increase in reproductive potential afterwards are likely to follow a non-senescent strategy, whereas species who mature more closely to their maximum attainable size are more likely to show senescence.

An exception is the special case of costly reproduction and cheap maintenance and growth. This is the only case where the level of extrinsic mortality can shift a strategy from a senescent to a non-senescent one. Here, the difference in vitality at maturity between two otherwise similar organism, one of which exhibits non-senescence and the other senescence, can be small.

In general, however, it can be stated that the the qualitative shape of a life history is determined by the returns to scale in growth and maintenance and reproduction. Senescence, i.e. exponentially increasing mortality during adult ages, is the prevalent optimal strategy only if both reproduction and maintenance are costly. If maintenance is cheap, then exponentially increasing mortality is not favored. In this case, Senescence is never optimal but instead Subsenescence can be optimal. If maintenance is costly but reproduction is cheap then exponentially increasing mortality is part of an optimal life history strategy. Ensenescence is optimal.

The degree of senescence and the age of and vitality at maturity are determined by the η parameters and the mortality parameters: Higher values of either of the η parameters (i.e. investment becomes increasingly costly and less efficient) are associated with a higher degree of senescence and larger vitality at maturity. Generally, low overall mortality favors low degrees of senescence. Higher values of extrinsic mortality are associated with a higher degree of senescence and smaller vitality at maturity. Higher values of intrinsic mortality, on the other hand can favor larger vitality at maturity as long as overall mortality does not rise to a level where again early maturation, i.e. small vitality is more favorable.

In sum, the crucial determinants of the degree of senescence, i.e. of the fraction of lifetime reproduction realized at ages when mortality increases, are the η parameters as well as the mortality parameters. The crucial determinants of whether a species follows a senescent or a non-senescent life history strategy are the returns to investment. Last but not least this chapter contributes to the discussion about the importance of fitness sensitivities to the evolution of age-patterns of mortality and fertility – these sensitivities are weights on the trade-offs that are balanced by evolution.

6

Directions for Research

6.1 Orientation

The models and analyses of the preceding chapters have shown that senescence is not inevitable. Much more research is needed to understand why in some species mortality increases after maturity while in others it does not. My results raise an important new question for aging research: when does senescence vs. sustenance evolve?

The study of the evolution of the age-patterns of mortality and fertility is still a wide-open field of research waiting for exploration. In this chapter, I summarize my thoughts about what I think are the most interesting and important parts of the field to explore next.

6.2 Direct Extensions of my Models

In this section I will describe several directions for research that follow naturally from my work described in this monograph.

6.2.1 Linking Burden & Optimization

Whether senescence is due to a burden of deleterious mutations or is a byproduct of optimization among trade-offs has been and still is the subject of intense discussion. In the following, I will outline my ideas of how linking both approaches could help to resolve the debate.

A Future Project

An interesting research project is to develop a model that combines mutational burden and optimization. Using the model developed in

Chap. 5, this could be done by including a mutational load term in the mortality function. Initially, the mutational load would be equal to zero. The optimal mortality and fertility patterns, given zero mutational load, can then be used to calculate selection pressure. The selection pressure plus assumptions about the magnitude and possible age-pattern of the mutation rate will determine the new mutational load. Given the new mutational load, a new round of optimization could be done. With this approach one would be able to analyze what proportion of mortality is due to optimization and what proportion is due to mutational load.

The question is whether the procedure converges. If decreasing selection pressure allows deleterious mutations to accumulate and leads to a rising mortality pattern, then the new selection pressure would fall more quickly, allowing still more mutations to accumulate. As more mutations accumulate, mortality would start rising earlier and earlier. Therefore, this model could shed light on whether the feedback loop between traits and evolution can lead to an unraveling of the life history.

6.2.2 Measurable Quantities and Testable Hypotheses

Evolutionary demographic theory is based on models. Each model is based on assumptions that simplify reality. A set of models can form a theory that illuminates a broad range of the real world because different simple models shed light on different, specific aspects of reality. Some models generate general insights, other models lead to testable hypothesis, and still other models make both contributions. I have developed simple models that contribute general insights to evolutionary demographic theory. In the future I plan to rethink and maybe reformulate my models to get a handle on measurable quantities to derive testable hypotheses from them.

This effort is a direct extension of my work. The parameters k, κ, and δ in the vitality model can be used to set a time and size scale. What is needed is an explicit expression that links vitality and size. This relationship hinges on knowledge about the rate at which cells are lost.

The η parameters capture the costs of maintenance and reproduction and have a major influence on the results of the model. Therefore it is important to understand how to root them in reality. What are the magnitudes of the η parameters for different types of species? This question has to be answered to generate testable hypotheses from the model.

6.2.3 The Diversity of Aging

My work described in this monograph can be viewed as a theoretical exploration of the inter-species diversity of aging, i.e., of how varied aging can be for different species and what factors determine whether a species' strategy involves sustenance or senescence. I believe that the models I developed can also be usefully applied, if somewhat refocused, to studies of intra-species variability of aging patterns resulting from environmental cues or conditions. A further research project that derives from my work is to develop this kind of application of my models.

As an example, consider the article by Mair et al. [116]. This group, from Linda Partridge's laboratory, explored two kinds of phenotypic plasticity of aging in genetically-identical lines of *Drosophila.* They manipulated diet and demonstrated that flies shifted to a restricted diet experienced, for the rest of their lives, the same trajectory of lower mortality as flies kept on the restricted diet all their lives. In terms of my vitality model, this effect is most simply explained by a shift in the parameter b. The "vitality" of a fly is unchanged by the dietary shift: the shift influences how vitality determines mortality. In contrast, Mair et al. [116] show that a reduction in temperature slows the pace of mortality increase with age. This effect can be captured in my model by a change in the deterioration parameter δ, or by a more general change in all the "scale" parameters k, κ, and δ or by a change in the strategy π. Which of these possibilities best captures reality? Collaborative theoretical and empirical research might answer this intriguing question.

A variety of other researchers, including James Carey, Thomas Johnson, James Curtsinger and Marc Tatar, have conducted laboratory studies of how some environmental change alters subsequent age-patterns of mortality and, in some cases, fertility. It may be possible to interpret the results of such studies in terms of changes in the parameters of my models – and this might shed light on mechanisms that underlie the phenotypic plasticity of aging.

6.2.4 The Characteristics of Senescent vs. Non-senescent Species

Species can be classified according to various characteristics. A challenging direction for future research is to identify the characteristics that distinguish senescent from non-senescent species.

The results of my models suggest that species with the capability of continued growth after the onset of reproduction are candidates for non-senescent life-history strategies. The results of my vitality-based model of Chap. 5 suggest that the costs of growth and maintenance and the costs of reproduction (as captured by parameters η_g and η_r) are the major determinants of senescence vs. sustenance. Developing a theory that identifies the relevant traits in reality that correspond to low vs. high values of η_g and η_r is a promising direction for future research opened up by my work.

In this regard, modularity might prove to be an important trait that can be associated with inexpensive growth and maintenance. The ability to reproduce clonally from segregated body parts, i.e. vegetative propagation, might prove to be associated with inexpensive reproduction. More thought is needed to come up with plausible hypotheses about characteristics that correspond to particular values of η_g and η_r.

A deep understanding of the returns to scale of investment to basic processes of life will shed light on why some species senesce and other species do not. Chap. 5 develops several ideas for exploration: Are there differences in returns to scale for asexual and sexual reproduction? Can the η parameters change with age, maybe being concave early in live and transitioning to convex later in live? Could different levels of organization, from the molecule over the cell to the whole organism be associated with different shapes of trade offs? Much more research is needed to answer all these questions.

The results of Chap. 5 further suggest that the parameters for intrinsic (i.e. state-dependent) and extrinsic (i.e. state-independent) mortality conditions (as captured by parameters b and c respectively) mainly influence the degree of senescence of a life history. What characteristics in a species determines the level of b and c? Williams [212] provides one hypothesis of how mortality conditions should influence the patterns of senescence. Research on this question has been done, e.g. by Ricklefs [164] and Ricklefs and Scheuerlein [165], but further research is needed to understand the influence of mortality components on the evolution of aging.

Mapping typologies of species into a typology of aging would be a major step towards understanding the evolution of senescence vs. sustenance. The simplest typology of aging would distinguish between species with strategies of senescence vs. sustenance. A more elaborate typology could be based on the five age-patterns of mortality and fertility discussed in Chap. 5. In addition to classifying species according to the shape of age-patterns of mortality and fertility, species could also be classified by their time scale: is life measured in hours, days, weeks, months, years, decades or centuries? Similarly, their size scale could be used: is size measured in nanometers, micrometers, millimeters, centimeters, decimeters or meters? As discussed above, other possible classifications could be growth mode, i.e. determinate vs. indeterminate growth, or the structure of the body plan, i.e. modular vs. non-modular structure.

6.2.5 Alternative Applications

The vitality model is a general model that could also shed light on other aspects of life that influence successful survival and reproduction. One important aspect is learning. If the single state variable in my models is interpreted as including the level of knowledge or cognitive ability, then the change in state can be due to learning or loss in cognitive ability. If more experience and knowledge imply a lower risk of death and more reproductive success, then my model can be extended to apply to the evolution of learning.

6.3 Other Modeling Extensions

In this section, I discuss several other directions for research for developing evolutionary demographic models. Over the past three years, there has been a spate of stimulating research in this area and I cite some pathbreaking recent advances.

6.3.1 Density Effects

Trees do not move. To live they need space to stand on. Therefore population density is crucial in a forest. If all patches are taken, no seedlings can establish themselves. This is true not only for trees in a forest but for many plants in many environments. In a recent working paper, Doncaster and Seymour [50] show that this density effect can explain the evolution of the great longevity of Bristlecone Pines. If seeds can only root themselves on a patch freed by the death of an adult, then longer lived trees have an evolutionary advantage. Their offspring will occupy the space opened by the death of the shorter lived trees whose offspring will not have found space to successfully establish themselves. The density effect favors the evolution of longevity.

Density is not only important for trees. The abundance of individuals in a population can significantly influence the evolution of life-history traits in general. If density effects play a role, then Lotka's intrinsic rate of population increase r is not an appropriate measure of fitness. Charlesworth [24] suggests using the number of individuals in the so called critical age group instead. Mylius and Diekmann [132] analyze what fitness measure to use, given the specific way density constrains population dynamics. The fitness measure used in the models I developed in Chaps. 4 and 5 is in accordance with the results of Mylius and Diekmann [132, p. 4]. In my models, density affects fertility via the multiplicative parameter φ for all ages equally.

Abrams [3] analyzes theoretically how extrinsic mortality should affect senescence, given different scenarios of density dependence. Williams [212] hypothesized that individuals living under more hazardous conditions should exhibit faster senescence and thereby lower survival than individuals living under more benign conditions. Abrams [3] shows that Williams' hypothesis will not always be valid if density effects alter population dynamics.

Density could also affect the optimal phenotype in a population. This could help to explain a puzzle recently noted by Resznick et al. [163]. They observe that, for a population of guppies living under two different mortality regimes in the wild, individuals from the high-risk environment show better survival when brought into the laboratory than individuals from the low-risk environment, contrary to Williams' hypothesis. However, because fewer individuals survive in the dangerous habitat, density is lower than under safer conditions. Therefore, the optimal high-risk phenotype develops when resources are more abundant, while the low-risk phenotype develops when resources are scarce. If more abundant resources allow for better growth and development and if this influences adult mortality, then the high-risk phenotype can be more robust than the low-risk phenotype. This density effect as well as several other possible explanations for the guppy-puzzle are discussed by Abrams [4].

Bronikowski and Promislow [15] emphasize that, depending on how senescence is defined and what kind of condition-dependent mortality is prevalent, different long-term effects on the evolution of senescence can be expected.

6.3.2 Intergenerational Transfers

Resources are scarce. Therefore, the age-trajectory of resources available to an individual over the life course constrains the evolution of optimal life histories. In this regard, resource flows among individuals are a crucial fitness component. The common fitness measures R and r do not include intergenerational transfers and, in particular, parental care. This can seriously distort results for species with significant periods of offspring dependence. Indeed, the degree of independence and the level of mortality at birth both reflect initial parental investment in offspring. From this perspective, size at birth relative to size at reproductive maturity is an important quantity. Lee [109] points out that the act of giving birth in itself can be interpreted as a transfer from mother to child. Therefore transfers should generally be captured by any measure of fitness.

Chu and Lee [36] and Robson and Kaplan [168] study conditions under which transfers from adult to offspring can be optimal: they model the co-evolution of longevity and transfers in human populations. Modeling efforts along these lines could explain the decline in mortality during development as well as the modest rather than steep increase at post-reproductive ages.

6.3.3 Environmental Fluctuations

Environmental fluctuations are certain over the life course of nearly all species. But their timing and magnitude can be highly uncertain. Natural selection needs time to work. If the environment changes faster than it takes selection to be effective, then chance plays a major role in favoring one species over another from one moment to the next. Populations can keep on fluctuating and might not reach a stable age-distribution. In variable environments the intrinsic rate of population increase is a poor measure of fitness because it assumes a stable population. Instead, the stochastic growth rate should be used to measure fitness (Orzack and Tuljapurkar [143], for a review see Tuljapurkar [199]). In a changing environment, the intrinsic rate of population increase r can be negative at every point in time but the stochastic growth rate can be positive: r does not capture real population dynamics.

Ripley and Caswell [166] demonstrate that an indicator of selection pressure – namely the relative change in the stochastic growth rate induced by changes in adult growth and survival of soft-shell clams – is strongly dependent on the amount of uncertainty in the recruitment of baby-clams. This state-dependent analysis implies that their indicator of selection pressure can increase with age if this uncertainty is large.

The development of phenotypes depends on the environment. Environmental cues can switch life histories between alternative age-trajectories of mortality and fertility most suitable to current conditions; some phenotypes can have prolonged life expectancy [7, 20, 63]. If life is harsh, nematode worms, for instance, can enter a state of very low metabolic activity, called the dauer state, that enables the worm to survive long periods of drought. Switching strategies require survival and reproductive patterns to be highly plastic.

In a recent issue of *Science*, Kussell and Leibler [104] offer a new method for approximating long-term reproductive success in fluctuating environments. Organisms can switch phenotypes according to the prevalent environment. Switching rates turn out to mimic the rate at which the environment is fluctuating. Furthermore, two extreme strategies of switching are compared – responsive vs. stochastic switching. Kussell and Leibler [104] show that switching strategies will be responsive or stochastic, depending on whether the costs of sensing the environment match the gains in reproductive success. An important determinant of this decision is the speed at which environments fluctuate. The information content of the environment (entropy) appears explicitly in the optimal solution, pointing to a deep connection between population biology and information theory.

6.3.4 Population Dynamics

It is useful to assume optimal equilibrium when studying whether non-senescence could be optimal at all. Research is needed to relax this assumption to better understand the domain of non-senescence vs. senescence. Given within-species dynamics like frequency dependence, could a non-senescent strategy be invaded by an alternative, senescent variant?

Survival is heavily influenced by the ability to resist diseases. A more or less costly immune system is necessary to fight the threats from the fast-evolving micro world. Given across-species dynamics like the co-evolution of the micro and macro world, how does the never-ending battle with parasites influence the evolution of senescence? More generally, some species are prey and other species are predators. Almost all species compete with other species for food and other resources. How does the competition among species influence age-patterns of mortality and fertility?

6.3.5 Summary

The models developed in this monograph were designed to shed light on whether non-senescent life-history strategies could be optimal. Further research can deepen and extend evolutionary demographic theory in various directions. In the previous sections I have highlighted the directions that I think are of most immediate interest and importance.

In particular, I laid out several research projects that directly derive from my work, namely:

- Integrating optimization and the burden of deleterious mutations in a single model,
- Reformulating my models such that the parameters are measurable and testable hypotheses can be derived,
- Focusing my models so that they can be used to understand how a species responds to changes in laboratory conditions, such as dietary or temperature manipulations,
- Mapping typologies of species into typologies of aging, and
- Applying the general model to alternative questions such as the co-evolution of longevity and learning.

In addition I have outlined five other directions for further evolutionary-demographic modeling, involving

- Density effects,
- Intergenerational transfers,
- Fluctuating environments,
- Intra-species population dynamics, and
- Inter-species population dynamics.

6.4 Prospects for Evolutionary Demography

Evolutionary Demography is an interdisciplinary area of research that has been newly evolving in recent years. In the following sections I highlight three lines along which the field could move forward. First, some canonical ideas need to be rethought. Second, new data, methods and measures are needed. Third, aging – the processes of change over age – can only be understood in the light of both senescence and sustenance together.

6.4.1 Moving Beyond the Burden of "Deleterious Fixations"

The phrase "deleterious fixations" is meant to emphasize that research on aging has been and still is influenced by long-held "truths" that channel thinking into directions that are limited and might even be wrong.

One of these fixations has successfully been rethought. For a long time, lifespan was believed to be strictly limited and specific to a species, i.e. nothing could be done about aging. The origin of the species-specific, limited lifespan paradigm can be traced back to Aristotle and Buffon. But over the past two decades gerontology has experienced a paradigm shift. Many experiments on flies, worms, yeast, rodents and other species led to the discovery that dietary restriction can prolong survival, helping shape the newly emerging insight that lifespan is not limited but plastic. Vaupel et al. [201] present age-patterns of mortality based on large sample sizes that do not increase steeply but instead level off or even decline at later ages for several species, thereby disproving the limited lifespan paradigm. Research on nematode worms, starting with Klass and Hirsch [99] and Johnson and Wood [89], demonstrates that changes in single genes can radically alter longevity.

One of the most remarkable examples of the plasticity of aging is presented in a paper by Mair et al. [116] that shows that changes in diet enable switching up and down between different mortality curves in *Drosophila*. Vaupel et al. [205] point out that similar patterns of switches have been observed in humans. Vaupel and colleagues show that mortality is plastic in humans even at advanced ages. One illustration is the convergence of mortality patterns in East and West Germany after reunification.

Indeed, a lot can be done about aging. The shift from the limited to the plastic lifespan paradigm is a major step forward in understanding senescence, exemplifying the importance of moving beyond a "deleterious fixation".

In the following, I list some other recalcitrant concepts that have channeled thinking on aging.

- Universal senescence
  Hamilton made the dogmatic claim that the force of selection inevitably declines, thus postulating the universality of senescence. This has restricted creative thinking about possible age-patterns of mortality.
  How universal is senescence?
- Gompertz Law
  It is widely believed that the age-pattern of mortality follows Gompertz law, but is it a law? We do not know what species exhibit this pattern over what range of age.
  How universal is an exponentially increasing hazard of death?
- No senescence in the wild
  It is often asserted that senescence is not experienced in the wild because individuals do not live long enough due to a high extrinsic hazard of death. This conjecture is intuitively appealing but it might

be wrong, as pointed out by Nesse [138] and Carey and Gruenfelder [18]. Carey and Gruenfelder summarize information available on the role of the elderly in primates, elephants and whales. Furthermore, Carey's recent observation of supine behavior in medflies – flies approaching death start lying on their backs, taking a rest once in a while over the remaining days of their lives – indicates that interesting patterns of senescence may be open to study.
Is there senescence in the wild? What are the age-patterns of mortality in the wild compared to those of creatures in captivity?

- Extrinsic hazard of death
Is it useful to distinguish between "extrinsic" and "intrinsic" hazards of death? The "intrinsic" hazard depends on age or, more generally, on an individual's state or condition. Are there "extrinsic" hazards that are independent of age or condition?
Extrinsic mortality is sometimes understood to be captured by the difference in mortality patterns of animals in the wild compared to patterns of those in captivity. However, animals kept in the zoo cannot pursue their natural behavior, for instance running long distances. The lack of exercise and of other behaviors performed in natural environments might distort mortality patterns in artificial habitats. Therefore, extrinsic mortality is not captured simply by the difference between mortality patterns in the wild and in captivity.
Probably most causes of death are condition-dependent. Natural catastrophes that kill all members of a group independently of condition could be seen as extrinsic risk, but such catastrophes may be rare.
What causes of death are truly condition-independent for a particular species?

6.4.2 The Need for Data, Methods, and Measures

Future theories of the evolution of aging should rest on scientific evidence. So far, the empirical evidence available on the age-trajectories of mortality and fertility for most species is based on small sample sizes [56, 159, 189, 215]. Meaningful age-patterns of demographic schedules, however, need to be based on large numbers of individuals, especially when studying senescence, because the size of the "interesting", later age-groups is progressively diminished by death. Vaupel [202] and Vaupel et al. [201] review the current empirical evidence of age-trajectories of mortality for species that are based on large sample sizes. These species include humans, *Drosophila*, medflies, three other species of fruit flies, a parasitoid wasp, the nematode worm *C. Elegans*, and yeast.

Serious study of the process of aging requires knowledge about actual patterns across a wide range of very different species. Biologists interested in different species collect a large amount of data on their particular species to answer their particular questions. It would be useful to obtain knowledge about what data are out there and whether people would be willing to contribute their data to a large database that allows for broad comparative studies of life-history patterns. A comparative study of the qualitative age-trajectories of mortality and fertility including candidates from the whole range of species with sufficiently large sample sizes is essential for developing theories of the evolution of aging.[1]

Methods need to be developed and applied that allow extraction of as much information as possible from the data available. Combining information from different data sources can lead to more conclusive results as emphasized by Anatoli Yashin and colleagues [217, 218]. An important step has recently been taken by James Carey and colleagues [130]: they developed a method for constructing life tables for captured cohorts of unknown age.[2] Their method circumvents the necessity to follow individuals longitudinally in the wild from birth onwards.

In addition to the strong need for new data and methods, it is important to develop a deeper understanding of how to measure senescence and sustenance. I suggest defining senescence as was discussed in Chap. 1: senescence occurs if but only if the relative change in mortality with age exceeds the relative change in fertility. I further suggest a general measure for the degree of senescence of a whole life history (Chap. 5): the fraction of life time reproduction that is realized at ages at which mortality increases. When gathering data to get comparative evidence it is essential to agree upon what is to be measured and compared.

---

[1] The Max Planck Institute for Demographic Research in Rostock has started a project headed by Alexander Scheuerlein to collect data-sets on patterns of mortality, fertility and growth for non-human species, in captivity and in the wild. A related research initiative will be coordinated at Duke University by Cliff Cunningham and will involve Jim Clark, James R. Carey and others. Other researchers, including Shripad Tuljapurkar and Steven Orzack, Susan Alberts and Tim Coulson, are also in the process of building databases on age-trajectories of mortality, fertility, and growth. The ISIS (International Species Information System) provides data for species kept in zoos; these data have recently been used to calculated comparative life tables for selected species of captive animals [100].

[2] This became necessary because the capture-recapture approach is not feasible for some species, including the medflies studied by Carey.

Even though the ultimate interest of evolutionary demography is focused on patterns over age, the deeper causal link is more than likely with stage and not age. Models should be based on stage and incorporate a biologically justified link from stage to age. Empirical observations and theoretical insights should be used to identify the crucial stage-variables that determine mortality and fertility patterns of a species. These variables need to be measured and included in the data sets.

6.4.3 A New Burning Question

A major and very important focus of research over the last decades has been testing which of the two leading theories, mutation accumulation vs. antagonistic pleiotropy, has more power to explain the evolution of senescence. Half a century after Medawar, Williams and Hamilton, evidence has been published both for and against mutation accumulation and antagonistic pleiotropy. The debate has still not been settled. Recent contributions include Charlesworth and Hughes [30], Charlesworth [29], Hughes et al. [84], Partridge and Barton [148], Partridge [147] and Steinsaltz et al. [187].

This monograph shows that senescence and sustenance are two sides of the process of aging. One cannot be deeply understood without the other. The new burning question that arises from my work is: when does senescence vs. sustenance evolve? An overarching theme that could guide theoretical and empirical work is: to what extent are age-schedules shaped by adaptive vs. non-adaptive processes? What I have done in this monograph is to broaden the focus from

- mutation accumulation vs. antagonistic pleiotropy to explain senescence to
- adaptive vs. non-adaptive theories to explain senescence vs. sustenance.

Medawar, Williams and Hamilton developed the basic ideas of the evolutionary theories of aging. The broadened focus suggested here allows us a wider perspective.

Adaptive

Adaptive theories explain aging as a byproduct of evolutionary optimization. Such theories are based on models of optimization constrained by trade-offs. Antagonistic pleiotropy and the disposable soma theory are adaptive theories of senescence. Senescence, which in itself is always a maladaptive process, is selected for because the trade-offs that constrain the life history are such that the benefits in fitness outweigh the costs due to senescence.

Reliability theory [65, 81, 108, 203] is another adaptive approach to explain senescence. Individuals are adapted to functioning over a sufficient period to guarantee the transmission of their genes. The subsequent senescent process is a byproduct that is determined by the preceding adaptive pattern.

If senescence and sustenance, i.e. aging, is explained by adaptive processes, then understanding is needed of the factors that have a strong impact on selection pressure vs. the factors that change selection pressure only slightly. Identification of the "strong forces" vs. "weak forces" of selection would provide a priority list of factors for understanding what shapes the age pattern of demographic schedules and its underlying variables[3]. For instance, Smith et al. [181, p. 1042, Fig. 5] show that environmental conditions can radically change stage-specific (and thereby age-specific) selection pressure. That is, the factor "environment" changes the importance of the different life-history transitions among states. This means that the variability of the environment is a strong force of selection. Note that, for this example (a threatened floodplain plant), selection pressure is highly state- but not age-dependent.

The list of valuable extensions to evolutionary demographic models given above in Sect. 6.3 is, likewise, a list of strong forces of selection, i.e. variability of the environment, density dependence, resource transfers, dynamics within and across species and probably more. Clearly, these components could interact with each other. Such interactions together with trade-offs among life-history traits at different ages can lead to dynamics that are not captured by the simple age-specific changes assumed in the indicators for the force of selection discussed in Chap. 2.

Non-adaptive

A non-adaptive theory is about what the force of selection cannot achieve. From the viewpoint of a non-adaptive theory, senescence exists because evolution is not strong enough to eradicate it. Sustenance, on the other hand, cannot be explained by non-adaptive theories. Only adaptive approaches have the potential to fully explain the aging process, while non-adaptive theory can partially account for the senescent

[3] I am grateful to Marc Tatar and Daniel Promislow for discussions about this.

side of the story. This indicates that adaptive approaches will be more powerful in explaining the aging process, although non-adaptive approaches could still play some role in explaining senescence.

The theory of mutation accumulation is a non-adaptive theory. The successive weakening of the force of selection for or against mutations implies that these mutations become increasingly neutral. Neutral theory explains the fate of a gene due to genetic drift and this drift strongly depends on population size. In this regard, assumptions about the time-horizon (infinite vs. finite) and the rate at which mutations occur at different ages are crucial to the conclusions from any mutation-accumulation model.

Over the last few years mounting knowledge about the human genome has been accumulating. Sufficient data are now available to check for age-specific gene-expression patterns in humans and also in other species such as *Drosophila*. It is possible to compare the fraction of individuals exhibiting neutral versus non-neutral mutations at young versus old ages. Evolutionary theories of senescence predict that falling selection pressure should make non-neutral mutations look more and more like neutral mutations as age increases. So if the fraction of individuals exhibiting non-neutral age-specific mutations becomes more similar over age to the fraction with neutral mutations, then this would be evidence for senescence being influenced by a non-adaptive process.

Evolution is constrained by phylogenetic history. A species can exhibit a non-adaptive age-pattern because a particular evolutionary path channels traits to a limited, possibly sub-optimal range. These phylogenetic channels could only be overcome in the very long run. So both mutation accumulation and phylogeny are non-adaptive forces shaping aging.

Further Thoughts

Creative thinking about alternative approaches to explain aging is needed. What are possible factors that shape the age-trajectories of mortality and fertility? Williams's hypothesis suggests the crucial importance of the extrinsic hazard of death. His hypothesis has been tested and evidence has been found for and against it. The contradictory evidence shows that this single factor is not enough to explain the pace of senescence. Williams identified one important variable that now needs to be put into perspective with other possible candidates.

What combination of these candidates leads to what qualitative age-pattern? In particular, when does senescence evolve and when sustenance? Clear, testable hypotheses need to be derived from theoretical models and empirical observations for what qualitative patterns of mortality and fertility are expected and when. My models in Chaps. 4 and 5 are a first systematic contribution to answering this question. My findings suggest that attention should be given to the costs of maintenance and reproduction.

An equally interesting and related question is how plastic the process of aging can be. For instance, studies of human twins have shown that the same genome can be associated with different patterns of senescence due to phenotypic plasticity. Only 25 % of the variation among humans in life expectancy can be attributed to genetic variation [79, 125]. So, how heterogeneous are species with respect to aging? What species have high plasticity, what species have low plasticity, what characteristics determine the degree of plasticity? Understanding the plasticity of senescence and sustenance would provide a strong tool in steering our own process of aging in the most advantageous way, i.e. towards a long and healthy life.

6.5 Conclusion

Senescence and sustenance are described by the age-trajectories of mortality and fertility. The age-trajectories of mortality and fertility are the fundamental demographic schedules: they determine the dynamics and structures of populations. In particular, they determine a population's genetic structure and size. Evolution can be viewed as change in genetic structure and size of populations over time. Changes in genetic structure lead to changes in age-trajectories. Therefore, evolution molds and is molded by demographic schedules of mortality and fertility. To understand the evolution of life it is crucial to study these schedules. Mortality and fertility are deeply interconnected with each other and in particular with the age-schedule of growth. The models developed in the previous chapters shed new theoretical light on the evolution of the age-schedules of mortality, fertility and growth and their interconnections.

My models suggest that a remarkable variety of patterns may be optimal under different circumstances. The limited empirical data available suggests that species may exhibit a rich diversity of age-schedules of mortality, fertility and growth. Current understanding of the biology of aging is largely based on laboratory studies of a restricted range of species. Getting reliable data on a wide variety of species is a crucial research need.

The evolutionary demographic theory of aging should aim at illuminating senescence vs. sustenance through the study of the age-patterns of mortality, fertility and growth. In particular, the research should explain why some species have a quickly or slowly increasing hazard of death and why others have a constant or falling hazard of death. The models I have developed are a first step towards gaining a deeper understanding of the evolution of senescence vs. sustenance. They lead to the general insight that the costs of maintenance and reproduction are the major determinants shaping these patterns.

In addition to exploring alternative qualitative patterns, evolutionary demographic theory should shed light on questions such as why some species live on short time scales and others on long ones, why some species grow large and others stay tiny and why some species produce numerous small progeny while others produce only few large progeny compared to adult body size. Thinking about scales of time and size could aid in the understanding of what kinds of species exhibit senescence vs. sustenance.

These species can be classified according to several characteristics. How such typologies map onto the typologies of senescence vs. sustenance will undoubtedly be a stimulating direction for future research.

Senescence is not inevitable. Life provides an alternative strategy: sustenance. Sustenance can theoretically be an optimal life-history strategy and is empirically observed for some species. Sustenance may be the strategy for a great many species in which mortality appears to fall or be constant over age, at least over an extended period of life after reproductive maturity. More extensive empirical evidence is needed for a broad range of species beyond humans, rodents, flies, nematodes and yeast. My thesis, the central insight of this work, is: to deeply understand why some species senesce, it is necessary to understand why other species do not.

A

Vitality Model - Appendix

A.1 Solving Differential Equations

Solving for λ_ψ

The Maximum Principle requires that

$$\dot{\lambda}_\psi = -\frac{dH}{d\psi}, \tag{A.1}$$

hence

$$\dot{\lambda}_\psi = -e^{-\phi}(1-\pi)^{\eta_r}\,\epsilon_\psi - \lambda_\psi\,(\pi^{\eta_g}\,\epsilon_\psi - \delta) - \lambda_\phi\,\mu_\psi\,. \tag{A.2}$$

Solving the differential equation leads to

$$\lambda_\psi(a) = \left(-\int_0^a g_\psi\, e^{\int_0^x f_\psi(s)\,ds}\,dx + A\right)\, e^{-\int_0^a f_\psi(s)\,ds} \tag{A.3}$$

with

$$g_\psi \equiv e^{-\phi}(1-\pi)^{\eta_r}\,\epsilon_\psi + \lambda_\phi\,\mu_\psi \tag{A.4}$$

capturing the change in fertility and in mortality with respect to a change in vitality, and with

$$f_\psi \equiv \pi^{\eta_g}\,\epsilon_\psi - \delta \tag{A.5}$$

capturing the change in growth with respect to a change in vitality. Note that the change in energy with respect to vitality ϵ_ψ is given by

$$\epsilon_\psi = 0.75\,k\,\psi^{-0.25} - \kappa\,, \tag{A.6}$$

being the derivative of (5.12) with respect to vitality.

Applying the transversality condition (5.20) in (A.3) one can solve for the constant A and find

$$\lambda_\psi(a) = \int_a^\infty g_\psi \, e^{\int_a^x f_\psi(s)\, ds} \, dx \,, \tag{A.7}$$

i.e.

$$\lambda_\psi(a) = \int_a^\infty \Big(e^{-\phi} (1-\pi)^{\eta_r} \, \epsilon_\psi + \lambda_\phi \, \mu_\psi \Big) \times e^{\int_a^x \pi^{\eta_g} \epsilon_\psi - \delta \, ds} \, dx \,. \tag{A.8}$$

The shadow price of vitality at age a is given by the associated cumulated changes in fertility and mortality over all remaining ages discounted by the corresponding cumulative changes in growth.

Solving for λ_ϕ

The Maximum Principle further requires that

$$\dot{\lambda}_\phi = -\frac{dH}{d\phi} \,, \tag{A.9}$$

hence

$$\dot{\lambda}_\phi = e^{-\phi} (1-\pi)^{\eta_r} \, \epsilon(\psi) \tag{A.10}$$

and thus

$$\lambda_\phi = \int_0^a e^{-\phi} (1-\pi)^{\eta_r} \, \epsilon(\psi) \, dx + C \,. \tag{A.11}$$

Again applying the transversality conditions (5.20) helps to solve for the constant C:

$$\lambda_\phi(a) = -\int_a^\infty e^{-\phi} (1-\pi)^{\eta_r} \, \epsilon(\psi) \, dx \,. \tag{A.12}$$

The shadow price of the cumulative hazard of death at age a is the negative value of remaining reproduction at age a, i.e. the penalty for having one unit higher cumulative hazard.

The expression in (A.12) can be substituted in (A.8) to yield the expression for $\lambda_\psi(a)$:

$$\lambda_\psi(a) = \int_a^\infty e^{\int_a^x \pi^{\eta_g} \epsilon_\psi - \delta \, ds} \times \Big(e^{-\phi} (1-\pi)^{\eta_r} \, \epsilon_\psi + \frac{b}{\psi^2} \int_x^\infty e^{-\phi} (1-\pi)^{\eta_r} \, \epsilon(\psi) \, d\tau \Big) \, dx \,. \tag{A.13}$$

The shadow price of vitality is given by the benefits of increasing reproduction due to higher vitality as well as the gains in remaining reproduction due to lower mortality, both weighted by the change in growth. As long as an increase in vitality leads to faster growth, this weight is above one (revaluating), if the increase in vitality leads to slower growth, then the weight is below one (devaluating).

Solving for Vitality

The differential equation in (5.14) can be solved substituting $z^4 = \psi$. After solving and re-substituting ψ, the equation for vitality is given by

$$\psi(a) = \left[\frac{k}{4} \int_0^a \pi^{\eta_g} \, e^{-\frac{\kappa}{4} \int_s^a \pi^{\eta_g} \, d\tau - \frac{\delta}{4}(a-s)} \, ds \right. \\ \left. + \, \psi(0)^{\frac{1}{4}} \, e^{-\frac{\kappa}{4} \int_0^a \pi^{\eta_g} \, d\tau - \frac{\delta}{4} a} \right]^4 \tag{A.14}$$

where $\psi(0)$ corresponds to vitality at age zero.

Note that for $\pi = 0$ expression (A.14) simplifies to

$$\psi(a) = \psi(0) \, e^{-\delta a} , \tag{A.15}$$

and for $\pi = 1$ to

$$\psi(a) = \left[\frac{k}{\kappa + \delta} - e^{-0.25 \, (\kappa + \delta) \, a} \left(\frac{k}{\kappa + \delta} - \psi(0)^{0.25}\right)\right]^4 . \tag{A.16}$$

A.2 Proof of Non-Existence of an Optimal Solution for a Special Case

Given that both η_r and η_g exceed one it can be proven that for the special case of constant mortality (i.e. $b = 0$) no optimal solution exists.

Proof From (5.19) it follows that

$$H(\pi = 0) = \lambda_\psi \left(\epsilon(\psi) - \delta \, \psi\right) + \lambda \, \mu(\psi) \tag{A.17}$$

and

$$H(\pi = 1) = e^{-\phi} \, \epsilon(\psi) - \lambda_\psi \, \delta \, \psi + \lambda \, \mu(\psi) \, . \tag{A.18}$$

Inserting those equations into the inequality $H(0) > H(1)$ and rearranging terms leads to

$$e^{\phi} \lambda_{\psi} < 1 . \tag{A.19}$$

The current value of the shadow price of vitality has to be smaller than one forever. Note that it is the current value of the shadow price of vitality, $\lambda^c_{\psi} \equiv e^{\phi} \lambda_{\psi}$, that matters for the optimal solution (see (5.22)).

For the special case of $\mu(\psi) = c$, i.e. $b = 0$, it can be shown that at some age a condition (A.19) will be violated: Since $\pi = 0$, (5.31) can be written as

$$\lambda_{\psi}(a) = \int_a^{\infty} e^{-c\,x} \epsilon_{\psi} e^{-\delta\,(x-a)} \, dx . \tag{A.20}$$

For constant mortality, condition (A.19) becomes $e^{c\,a} \lambda < 1$. Thus, multiplying (A.20) by $e^{c\,a}$ yields

$$\lambda^c_{\psi}(a) = \int_a^{\infty} e^{-c\,(x-a)} \epsilon_{\psi} e^{-\delta\,(x-a)} \, dx . \tag{A.21}$$

Taking into account that for $\pi = 0$ vitality is given by (A.15) and energy changes with respect to vitality according to (A.6), (A.21) becomes

$$\begin{aligned} \lambda^c_{\psi}(a) &= \int_a^{\infty} \epsilon_{\psi} e^{-(c+\delta)\,(x-a)} \, dx \\ &= 0.75\,\psi(a)^{-0.25} \frac{k}{c + 0.75\,\delta} - \frac{\kappa}{c+\delta} . \end{aligned} \tag{A.22}$$

Does (A.19) hold? Inserting (A.22) and rearranging terms yields

$$\psi(a) \geq \left(\frac{0.75\,k\,(c+\delta)}{(\kappa + c + \delta)\,(c + 0.75\,\delta)} \right)^4 . \tag{A.23}$$

Since zero investment in growth ($\pi = 0$) causes vitality $\psi(a)$ to approach zero as age a approaches infinity while the right-hand side of (A.23) is a positive constant, condition (A.23) and thus condition (A.19) will definitely be violated at a certain low level of vitality.

A.3 The Algorithm

To solve the dynamic optimization problem I applied a dynamic programming approach by developing an algorithm following a backward procedure and assuming stepwise constant vitality [12]. Crucial to Bellman's approach is that the optimal decision does not depend on the past, but is based solely on the current state. The state determines possible current and future payoffs. An essential requirement for this backward optimization to work is the knowledge of an ultimate state with known payoffs, the ultimate future expectation. The procedure starts at this ultimate state and then works backwards along the state trajectory. If the mode of change can switch back and forth between growth and shrinkage, then such an ultimate state cannot be identified and the problem becomes intractable with Bellman's approach. My model constraint implies that the switch can only occur once. Since life necessarily starts off with growth, the switch is initially in up mode and can optionally change into down mode.

The state trajectory is assumed to be stepwise constant. The time it takes to change from vitality ψ to vitality $\psi \pm \Delta$ ($\Delta > 0$, step size) is given by the step time

$$\tau(\psi, \pi) = \frac{\Delta}{\dot{\psi}}, \tag{A.24}$$

where $\dot{\psi}$ is defined in Equation 5.4. Note that if vitality falls, then $\tau(\psi, \pi) = -\Delta/\dot{\psi}$ and if vitality is maintained then $\tau(\psi, \pi) = \infty$.

At each level of vitality the algorithm maximizes remaining reproduction, given by

$$R(\psi) = \int_0^{\tau} e^{-\mu(\psi)\, a}\, m(\psi, \pi)\, da + e^{-\mu(\psi)\, \tau(\psi,\pi)}\, R(\psi_{next}). \tag{A.25}$$

Since vitality is constant over the time interval τ, the integral in Equation A.25 can be solved, yielding

$$R(\psi) = \frac{m(\psi, \pi)}{\mu(\psi)} \left[1 - e^{-\mu(\psi)\tau(\psi,\pi)}\right] + e^{-\mu(\psi)\, \tau(\psi,\pi)}\, R(\psi_{next}). \tag{A.26}$$

Remaining reproduction is given by current reproduction weighted by the chance of dying in that interval and remaining reproduction at the subsequent level of vitality weighted by the probability of surviving the time interval.

The algorithm to determine the optimal investment trajectory $\pi^*(\psi)$ (the star indicates "optimal") has two parts, one for each mode. For this application, the ultimate state corresponds to a vitality of $\psi = 0$ and therefore to a mortality that is infinite and remaining reproduction of zero. Consequently, the first part of the algorithm begins in down mode at the end of possible state trajectories, i.e. at the last level of vitality $\psi > 0$ when the switch is in down mode. Since initial vitality equals one, it is convenient to choose $\psi = 1$. Then, the initial step is to find $\pi_d^*(1)$ and the corresponding $R_d^*(1)$ (the d indicates "down mode") using Equation A.26:

$$\begin{aligned}
\pi_d^*(1)) &= \max_{\pi \, \epsilon \, [0, \, \pi_0]} \; R_d(1) \qquad \text{(A.27)} \\
&= \max_{\pi \, \epsilon \, [0, \, \pi_0]} \; \frac{m(1, \pi)}{\mu(1)} \left[1 \; - \; e^{-\mu(1)\tau(1,\pi)}\right] \; + \; 0 \\
&= \max_{\pi \, \epsilon \, [0, \, \pi_0]} \; \frac{(1 \; - \; \pi)^{\eta_r}\,(k - \kappa)}{b + c} \\
&\qquad \times \left[1 \; - \; e^{-(b+c)\,\Delta\,/\,(\pi^{\eta_g}\,(k-\kappa)\,-\,\delta)}\right].
\end{aligned}$$

Note that my constraint implies that optimal investment π will lie between zero and π_0.

The procedure is repeated working backwards for all levels of vitality up to the maximum attainable vitality $\psi = \Psi$, determined by Equation 5.8. For each level of vitality the optimal investment is found by

$$\begin{aligned}
\pi_d^*(\psi) = \max_{\pi \epsilon [0, \, \pi_0]} \; & \frac{m(\psi, \pi)}{\mu(\psi)} \left[1 \; - \; e^{-\mu(\psi)\tau(\psi,\pi)}\right] \\
& + e^{-\mu(\psi)\,\tau(\psi,\pi)} \; R_d^*(\psi - \Delta).
\end{aligned} \qquad \text{(A.28)}$$

This part of the algorithm gives an optimal decision for each level of vitality in down mode.

Maximum attainable vitality Ψ gives the ultimate state for the second part of the algorithm. If the switch is in up mode and vitality is at its maximum attainable level Ψ, then the decision is whether to either stay in up mode and maintain maximum vitality or to switch into down mode and follow the already calculated optimal investment in down mode:

$$\pi_u^*(\Psi) \; = \; \begin{cases} \pi_0(\Psi) \text{ if } \quad R_u^*(\Psi) \; = \; \frac{m(\Psi, \pi_0)}{\mu(\Psi)} > R_d^*(\Psi) \\ \\ \pi_d^*(\Psi) \text{ otherwise.} \end{cases} \qquad \text{(A.29)}$$

Note that if mortality μ and fertility m are constant, then remaining reproduction is given by m/μ.

Then vitality is followed backwards, down to the smallest level of vitality $\psi = 1$. At each level of vitality the optimal investment is found by

$$\begin{aligned}\pi_u^*(\psi) &= \max_{\pi\epsilon[\pi_0,\,1]} R_u(\psi) \\ &= \max_{\pi\epsilon[\pi_0,\,1]} \frac{m(\psi,\pi)}{\mu(\psi)}\,(1 - e^{-\mu(\psi)\tau(\psi,\pi)}) \\ &\qquad + e^{-\mu(\psi)\,\tau(\psi,\pi)}\, R_u^*(\psi+\Delta)\end{aligned} \tag{A.30}$$

if $R_u^*(\psi) > R_d^*(\psi)$ and otherwise $\pi_u^*(\psi) = \pi_d^*(\psi)$. The second part of the algorithm gives an optimal strategy for each level of vitality in up mode.

The optimal strategy over the life course can be found by connecting the results from part one and two of the algorithm in the following way: Results are saved in the form of a vector

$$\begin{pmatrix} \textit{remaining reproduction} \\ \textit{mode of change} \\ \textit{vitality} \\ \textit{investment} \\ \textit{time} \end{pmatrix} = \begin{pmatrix} R^*(\psi) \\ G,\ S \textit{ or } M \\ \psi \\ \pi^*(\psi) \\ \tau^*(\psi) \end{pmatrix} \tag{A.31}$$

Note that the variable "mode of change" takes on the value G for growth if vitality increases, S for shrinkage if vitality decreases and M for maintenance if vitality remains constant. For each level of vitality, the optimal vector is saved in a list. The optimal solution can be found from this list by connecting the vectors in the right order. The only logical succession of vectors regarding the mode of change are $(G,\ldots,G,M)$, $(G,\ldots,G,S,\ldots,S,M)$ and $(G,\ldots,G,S,\ldots,S)$. Trivially, vectors need be be nested according to subsequent levels of vitality.

Finally, the constant parameter φ can be used to adjust R^* to be equal to one. This implies that density effects produce population stationarity by reducing life-time fertility [24, 132].

References

[1] P. Aberg. A demographic study of two populations of the seaweed *Ascophyllum nodosum*. *Ecology*, 73(4):1473–1487, 1992.

[2] Peter A. Abrams. The fitness costs of senescence: The evolutionary importance of events in early adult life. *Evolutionary Ecology*, 5:343–360, 1991.

[3] Peter A. Abrams. Does increased mortality favor the evolution of more rapid scenescence? *Evolution*, 47(3):877–887, 1993.

[4] Peter A. Abrams. Mortality and lifespan. *Nature*, 431:1048–1049, 2004.

[5] Peter A. Abrams and D. Ludwig. Optimality theory, Gompertz' law, and the disposable soma theory of senescence. *Evolution*, 49(6):1055–1066, 1995.

[6] E.R. Alvarez-Buylla and M. Martinez-Ramos. Demography and allometry of *Cecropia obtusifolia*, a neotropical pioneer tree - an evaluation of the climax pioneer paradigm for tropical rain forests. *Journal of Ecology*, 80(2):275–290, 1992.

[7] G.V. Amdam and S.W Omholt. The regulatory anatomy of honeybee lifespan. *Journal of Theoretical Biology*, 216(2):209–228, 2002.

[8] James J. Anderson. A vitality-based model relating stressors and environmental properties to organism survival. *Ecological Monographs*, 70(3):445470, 2000.

[9] Robert Arking, Stephen Buck, Angela Berrios, Susan Dwyer, and George T. III Baker. Elevated paraquat resistance can be used as a bioassay for longevity in a genetically based long-lived strain of *Drosophila*. *Developmental Genetics*, 12:362–370, 1991.

[10] R.C. Babcook. Comparative demography of three species of scleractinian corals using age- and size-dependent classifications. *Ecological Monographs*, 61(3):225–244, 1991.

[11] Annette Baudisch. Hamilton's indicators of the force of selection. *Proceedings of the National Academy of Sciences, USA*, 102(23):8263–8268, 2005.

[12] Richard E. Bellman. *Dynamic Programming.* New Jersey : Princeton University Press, 1965.

[13] G. E. P. Box. Robustness in scientific model building. In R. L. Launer and G. N. (Eds.) Wilkinson, editors, *Robustness in statistics*, pages 201–236. New York: Academic Press., 1979.

[14] A. M. Bronikowski, Altmann J. Alberts, S. C., C. Packer, K.D. Carey, and M. Tatar. The aging baboon: Comparative demography in a non-human primate. *Proceedings of the National Academy of Sciences, USA*, 99:9591–9595, 2002.

[15] Anne M. Bronikowski and Daniel E.L. Promislow. Testing evolutionary theories of aging in wild populations. *TRENDS in Ecology and Evolution*, 20(6):271–273, 2005.

[16] Michael Bulmer. *Theoretical Evolutionary Ecology.* Oxford: Freeman, 1995.

[17] R. Bürger. *The mathematical theory of selection, recombination, and mutation.* John Wiley, Chichester, New York, 2000.

[18] James R. Carey and C.A. Gruenfelder. Population biology of the elderly. In Kenneth W. Wachter and Caleb E. Finch, editors, *Between Zeus and the Salmon*, pages 127–160, Washington, D.C., 1997. National Academy Press.

[19] James R. Carey and Debra S. Judge. *Longevity records : Life spans of mammals, birds, amphibians, reptiles, and fish.* Odense University Press, 2000.

[20] James R. Carey, Pablo Liedo, Hans Georg Müller, Jane Ling Wang, and James W. Vaupel. Dual modes of aging in mediterranean fruit fly females. *Science*, 281:996–998, 1998.

[21] James R. Carey, Pablo Liedo, Dina Orozco, and James W. Vaupel. Slowing of mortality rates at older ages in large medfly cohorts. *Science*, 258(5081):457–461, 1992.

[22] Hal Caswell. On the equivalence of maximizing reproductive value and maximizing fitness. *Ecology*, 61(1):19–24, 1980.

[23] Hal Caswell. *Matrix population models : Construction, analysis, and interpretation.* Sunderland, Sinauer, Massachusetts, 2nd edition, 2001.

[24] Brian Charlesworth. Selection in populations with overlapping generations. V. Natural selection and life histories. *American Naturalist*, 107(954):303–311, 1973.

[25] Brian Charlesworth. Selection in populations with overlapping generations. VI. Rates of gene frequency and population growth rate. *Theoretical Population Biology*, 6:108–133, 1974.

[26] Brian Charlesworth. Optimization models, quantitative genetics, and mutation. *Evolution*, 44(3):520–538, 1990.

[27] Brian Charlesworth. *Evolution in age-structured populations*. Cambridge University Press, Cambridge, 2nd edition, 1994.

[28] Brian Charlesworth. Perspectives. *Genetics*, 156:927–931, 2000.

[29] Brian Charlesworth. Patterns of age-specific means and genetic variance of mortality rates predicted by the mutation-accumulation theory of ageing. *Journal of theoretical Biology*, 210:47–65, 2001.

[30] Brian Charlesworth and Kimberly A. Hughes. Age-specific inbreeding depression and components of genetic variance in relation to the evolution of senescence. *Proceedings of the National Academy of Sciences, USA*, 93:61406145,, 1996.

[31] Brian Charlesworth and Jesus Alberto Leon. The relation of reproductive effort to age. *American Naturalist*, 110(973):449–459, 1976.

[32] Brian Charlesworth and Linda Partridge. Ageing: Levelling of the Grim Reaper. *Current Biology*, 7:R440–R442, 1997.

[33] Eric L. Charnov. *Life History Invariants*. Oxford Series in Ecology and Evolution. Oxford University Press, 1993.

[34] Eric L. Charnov and James F. Gillooly. Size and temperature in the evolution of fish life histories. *Integrative and Comparative Biology*, 44:494–497, 2004.

[35] Eric L. Charnov, T.F. Turner, and K.O. Winemiller. Reproductive constraints and the evolution of life histories with indeterminate growth. *Proceedings of the National Academy of Sciences, USA*, 98(16):9460–9464, 2001.

[36] C. Y. Cyrus Chu and Ronald D. Lee. The co-evolution of intergenerational transfers and longevity: An optimal life history approach. *Theoretical Population Biology*, 69:193201, 2006.

[37] Mariusz Cichon. Evolution of longevity through optimal resource allocation. *Proceedings of the Royal Society London*, B 264:1383–1388, 1997.

[38] Mariusz Cichon. Diversity of age-specific reproductive rates may result from ageing and optimal resource allocation. *Journal of Evolutionary Biology*, 14:180–185, 2001.

[39] Mariusz Cichon and Jan Kozlowski. Ageing and typical survivorship curves result from optimal resource allocation. *Evolutionary Ecology Research*, 2:857–870, 2000.

[40] David J. Clancy, David Gems, Lawrence G. Harshman, Sean Oldham, Hugo Stocker, Ernst Hafen, Sally J. Leevers, and Linda Partridge. Extension of life-span by loss of CHICO, a *Drosophila* insulin receptor substrate protein. *Science*, 292:104 – 106, 2001.

[41] C.W. Clark and M. Mangel. *Dynamic State Variable Models in Ecology: Methods and Applications.* Oxford University Press, 2000.

[42] Lamont C. Cole. The population consequences of life history phenomena. *Quarterly Review of Biology*, 29(2):103–137, 1954.

[43] Alexander Comfort. *The biology of senescence.* Routledge and Kegan Paul, 1956.

[44] R.E. Cook. Germination and size-dependent mortality in *Viola blanda. Oecologia*, 47:115–117, 1980.

[45] D. R. Cox. Regression models and life tables. *Royal Statistical Society.*, 34:187–220, 1972.

[46] J. F. Crow and M. Kimura. *An introduction to population genetics theory.* Harper and Row, New York, 1970.

[47] J.W. Curtsinger, H.H. Fukui, D.R. Townsend, and J.W. Vaupel. Demography of genotypes: failure of the limited life-span paradigm in drosophila melanogaster. *Science*, 258:461–463, 1992.

[48] Charles Darwin. *On the Origin of Species. A Facsimile of the First Edition.* Harvard University Press, Cambridge, 1964.

[49] Gerdien De Jong. Is invariance across animal species just an illusion? *Science*, 309:1193–1195, 2005.

[50] C. P. Doncaster and Robert M. Seymour. Fitness for ever longer reproductive life. unpublished, 2005.

[51] J. W. Drake, B. Charlesworth, D. Charlesworth, and J. F. Crow. Rates of spontaneous mutation. *Genetics*, 148:1667–1686, 1998.

[52] F. Drenos and T.B.L. Kirkwood. Modelling the disposable soma theory of ageing. *Mechanisms of Ageing and Development*, 126:99–103, 2005.

[53] Thomas A. Ebert and John R. Southon. Red sea urchins (*Strongylocentrotus franciscanus*) can live over 100 years: confirmation with A-bomb ^{14}carbon. *Fishery Bulletin*, 101(4):915–922, 2003.

[54] W. J. Ewens. *Mathematical population genetics.* Springer Verlag, New York, Heidelberg, Berlin, 1979.

[55] A. Eyre-Walker and P. D. Keightley. High genomic deleterious mutation rates in hominids. *Nature*, 397:344–347, 1999.

[56] Caleb E. Finch. *Longevity, Senescence, and the Genome.* University of Chicago Press, Chicago, 1990.

[57] C.E. Finch. Variations in senescence and longevity include the possibility of negligible senescence. *Journal of Gerontology: Biological Sciences*, 53A(4):B235–B239, 1998.

[58] C.E. Finch and Steven N. Austad. History and prospects: Symposium on organisms with slow aging. *Experimental Gerontology*, 36:593–597, 2001.

[59] R.A. Fisher. The fundamental theorem of natural selection. In *The Genetical Theory of Natural Selection*, pages 25–30. Clarendon Press, Oxford, 1930. (Reprinted and revised, 1958. New York: Dover).

[60] Kevin Flurkey, John Papaconstantinou, Richard A. Miller, and David E. Harrison. Life-span extension and delayed immune and collagen aging in mutant mice with defects in growth hormone production. *Proceedings of the National Academy of Sciences, USA*, 98:6736–6741, 2001.

[61] R. M. Fragen. An optimal life history in which reproductive effort decreases with age. *American Naturalist*, 106:258–261, 1972.

[62] Madhav Gadgil and William H. Bossert. Life historical consequences of natural selection. *American Naturalist*, 104(935):1–24, 1970.

[63] Michael P. Gardner, David Gems, and Mark E. Viney. Extraordinary plasticity of ageing in a nematode. submitted, 2006.

[64] Shea N. Gardner and Marc Mangel. When can a clonal organism escape senescence? *American Naturalist*, 150(4):462–490, 1997.

[65] L.A. Gavrilov and N.S. Gavrilova. *The Biology of Life Span.* Harwood Academic Publishers, Chur, Switzerland, 1991.

[66] L.U. Gerdes, B. Jeune, K. Andersen-Ranberg, H. Nybo, and J. W. Vaupel. Estimation of Apolipoprotein E genotype-specific relative mortality risk from the distribution of genotypes in centenarians and middle-aged men: Apolipoprotein E gene is a “frailty gene,” not a "longevity gene”. *Genetic Epidemiology*, 19:202–210, 2000.

[67] Tamara R. Golden and Simon Melov. Microarray analysis of gene expression with age in individual nematodes. *Aging Cell*, 3:111, 2004.

[68] T. Good and M. Tatar. Age-specific mortality and reproduction respond to adult dietary restriction in *Drosophila melanogaster*. *Journal Insect Physiology*, 47:1467–1473, 2001.

[69] Daniel Goodman. Optimal life histories, optimal notation, and the value of reproductive value. *American Nationalist*, 119(6):803–823, 1982.

[70] R.W. Grigg. Population dynamics of two gorgonian corals. *Ecology*, 58(2):278–290, 1977.

[71] N.M. Guedje, J. Lejoly, B. A. Nkongmeneck, and W.B.J. Jonkers. Population dynamics of *Garcinia lucida* (clusiaceae) in cameroonian atlantic forests. *Forest Ecology and Management*, 177(1-3):231–241, 2003.

[72] J. B. S. Haldane. A mathematical theory of natural and artificial selection, IV. *Proceedings of Cambridge Philosopical Society*, 23:607–615, 1927.

[73] J. B. S. Haldane. The effect of variation in fitness. *American Naturalist*, 71:337–349, 1937.

[74] J. B. S. Haldane. The cost of natural selection. *Journal of Genetics*, 55:511–524, 1957.

[75] W. D. Hamilton. The moulding of senescence by natural selection. *Journal of Theoretical Biology*, 12:12–45, 1966.

[76] W. D. Hamilton. *Narrow Roads of Gene Land: The Collected Papers of W.D. Hamilton*, volume 1: Evolution of Social Behaviour. W.H. Freeman Spektrum, Oxford, New York, Heidelberg, 1996.

[77] J.L. Harper. *Population biology of plants*. Academic Press, New York, New York, USA, 1977.

[78] A. K. Hegazy. Age-specific survival, mortality and reproduction, and prospects for conservation of *Limonium delicatulum*. *Journal of Applied Ecology*, 29(3):549–557, 1992.

[79] A.M. Herskind, Matt McGue, Niels V. Holm, T.I.A. Soerensen, Bent Harvald, and James W. Vaupel. The heritability of human longevity: A population-based study of 2872 Danish twin pairs born 1870-1900. *Human Genetics*, 97:319–323, 1996.

[80] B. Heulin, K. Osenegg-Leconte, and D. Michel. Demography of a bimodal reproductive species of lizard (*Lacerta vivipara*): Survival and density characteristics of oviparous populations. *Herpetologica*, 53(4):432–444, 1997.

[81] Shiro Horiuchi. Interspecies differences in the life span distribution: Humans versus invertebrates. In James R. Carey and Shripad Tuljapurkar, editors, *Life Span. Evolutionary, Ecological,*

and Demographic Perspectives, pages 127–151, New York, 2003. Population Council.

[82] K.A. Hughes. The evolutionary genetics of male life-history characters in *Drosophila melanogaster*. *Evolution*, 49:521–537, 1995.

[83] K.A. Hughes and Brian Charlesworth. A genetic analysis of senescence in *Drosophila*. *Nature*, 367:64–66, 1994.

[84] Kimberly A. Hughes, Julie A. Alipaz, Jenny M. Drnevich, and Rose M. Reynolds. A test of evolutionary theories of aging. *Proceedings of the National Academy of Sciences, USA*, 99:1428614291, 2002.

[85] R.N. Hughes and D.J. Roberts. Comparative demography of *Littorina rudis, L. nigrolineata* and *L. neritoides* on three contrasted shores in north wales. *Journal of Animal Ecology*, 50(1):251–268, 1981.

[86] T.P. Hughes and J.B.C. Jackson. Population dynamics and life histories of foliaceous corals. *Ecological Monographs*, 55:141–166, 1985.

[87] D. S. Hwangbo, B. Gershman, M.P. Tu, M. Palmer, and M. Tatar. *Drosophila* dFOXO controls lifespan and regulates insulin signalling in brain and fat body. *Nature*, 429:562–566, 2004.

[88] J.A. Jimenes and A.E. Lugo. Tree mortality in mangrove forests. *Biotropica*, 17:177–185, 1985.

[89] T. E. Johnson and W. B. Wood. Genetic analysis of life-span in *Caenorhabditis elegans*. *Proceedings of the National Academy of Sciences, USA*, 79:6603–6607, 1982.

[90] Morton I. Kamien and Nancy L. Schwartz. *Dynamic Optimization: The Calculus of Variations and Optimal Control in Economics and Management*. North Holland, Amsterdam, New York, London, Tokyo, second edition, 1991.

[91] Hillard S. Kaplan and Arthur J. Robson. The emergence of humans: The coevolution of intelligence and longevity with intergenerational transfers. *Proceedings of National Academy of Science, USA*, 99(15):10221–10226, 2002.

[92] Peter D. Keightley and Brian Charlesworth. Genetic instability of *C. Elegans* comes naturally. *TRENDS in Genetics*, 21(2):67–70, 2005.

[93] J.A. Kemperman and B.V. Barnes. Clone size in American aspens. *Canadian Journal of Botany*, 54:2603–2607, 1976.

[94] Nathan Keyfitz. *Introduction to the Mathematics of Population*. Addison Wesley, Reading, 1968.

[95] M. Kimura and T. Maruyama. The mutational load with epistatic gene interaction in fitness. *Genetics*, 54:1337–51, 1966.

[96] J. F. C. Kingman. *Mathematics of genetic diversity.* Society for industrial and applied mathematics, 1980.

[97] T. B. L. Kirkwood. *Physiological Ecology: An Evolutionary Approach to Resource Use*, chapter Repair and its Evolution: Survival Versus Reproduction, pages 165–189. Blackwell Scientific, Oxford, 1981.

[98] T.B.L. Kirkwood. Evolution of ageing. *Nature*, 270:301–304, 1977.

[99] Michael R. Klass and David Hirsch. Non-aging developmental variant of *Caenorhabditis elegans*. *Nature*, 260:523–525, 1976.

[100] Iliana V. Kohler, Samuel H. Preston, and Laurie Bingaman Lackey. Comparative mortality levels among selected species of captive animals. *Demographic Research*, 15:413–434, 2006.

[101] S. A. L. M. Koojiman. *Dynamic energy and mass budgets in biological systems.* Cambridge University Press, 2nd edition, 2000.

[102] Jan Kozlowski. Optimal allocation of resources explains interspecific life-history patterns in animals with indeterminate growth. *Proceedings of the Royal Society London*, B 263:559–566, 1996.

[103] Jan Kozlowski and Anatoly T. Teriokhin. Allocation of energy between growth and reproduction: The Pontryagin Maximum Principle solution for the case of age- and season-dependent mortality. *Evolutionary Ecology Research*, 1:413–441, 1999.

[104] Edo Kussell and Stanislas Leibler. Phenotypic diversity, population growth, and information in fluctuating environments. *Science*, 309:2075–2078, 2005.

[105] Russell Lande. A quantitative genetic theory of life history evolution. *Ecology*, 63(3):607–615, 1982.

[106] Gary N. Landis, Diana Abdueva, Dmitriy Skvortsov, Junsheng Yang, Beth E. Rabin, James Carrick, Simon Tavar, and John Tower. Similar gene expression patterns characterize aging and oxidative stress in *Drosophila melanogaster*. *Proceedings of the National Academy of Sciences, USA*, 101:7663–7668, 2004.

[107] D. M. Lavigne. Similarity in energy budgets of animal populations. *Journal of Animal Ecology*, 51:195–206, 1982.

[108] Herve LeBras. Lois de mortalite a age limite. *Population*, 31:655–692, 1976.

[109] Ron Lee. Rethinking the evolutionary theory of aging: Transfers, not births, shape senescence in social species. *Proceedings of the*

National Academy of Sciences of the United States of America, 100(16):9637–9642, 2003.

[110] Jeff Leips and Trudy F. C. Mackay. Quantitative trait loci for life span in drosophila melanogaster: Interactions with genetic background and larval density. *Genetics*, 155:1773–1788, 2000.

[111] Jesus Alberto Leon. Life histories as adaptive strategies. *Journal of Theoretical Biology*, 60:301–335, 1976.

[112] Yi-Jyun Lin, Laurent Seroude, and Seymour Benzer. Extended life-span and stress resistance in the *Drosophila* mutant Methuselah. *Science*, 282:943 – 946, 1998.

[113] G. J. Lithgow, T. M. White, S. Melov, and T. E. Johnson. Thermotolerance and extended life-span conferred by single-gene mutations and induced by thermal stress. *Proceedings of the National Academy of Sciences, USA*, 92:7540–7544, 1995.

[114] Daniel Lonard and Ngo Van Long. *Optimal Control and Static Optimization in Economics.* Cambridge University Press, Cambridge, 1992.

[115] Alfred J. Lotka. *Elements of Mathematical Biology.* Reprinted 1956 by Dover Publications, Inc., New York, 1924.

[116] William Mair, Patrick Goymer, Scott.D Pletcher, and Linda Partridge. Demography of dietary restriction and death in *Drosophila. Science*, 301:1731–1733, 2003.

[117] Olvi L. Mangasarian. Sufficient conditions for the optimal control of nonlinear systems. *SIAM Journal on Control and Optimization*, 4:139–152, 1966.

[118] M. Mangel and C.W. Clark. *Dynamic Modeling in Behavioral Ecology.* Princeton University Press, 1988.

[119] Marc Mangel. Oviposition site selection and clutch size in insects. *Journal of Mathematical Biology*, 25:1–22, 1987.

[120] Marc Mangel and Michael B. Bonsall. The shape of things to come: Using models with physiological structure to predict mortality trajectories. *Theoretical Population Biology*, 65:353359, 2004.

[121] Marc Mangel and Stephan B. Munch. A life-history perspective on short- and long-term consequences of compensatory growth. *American Naturalist*, 166, 2005.

[122] Marc Mangel and Judy Stamps. Trade-offs between growth and mortality and the maintenance of individual variation in growth. *Evolutionary Ecology Research*, 3:583–593, 2001.

[123] Daniel E. Martinez. Mortality patterns suggest lack of senescence in Hydra. *Experimental Gerontology*, 33(3):217–225, 1998.

[124] Steven A. McCarroll, Coleen T. Murphy, Sige Zou, Scott D. Pletcher, Chen-Shan Chin, Yuh Nung Jan, Cynthia Kenyon, Cornelia I. Bargmann, and Hao Li. Comparing genomic expression patterns across species identifies shared transcriptional profile in aging. *Nature Genetics*, 36:197 – 204, 2004.

[125] Matt McGue, James W. Vaupel, Niels Holm, and Bent Harvald. Longevity is moderately heritable in a sample of Danish twins born 1870-1880. *Journal of Gerontology, Biological Sciences*, 48(6):B237–B244, 1993.

[126] P. B. Medawar. *Uniqueness of the Individual*, chapter An Unsolved Problem of Biology, pages 44–70. H.K. Lewis, 1952.

[127] P. A. P. Moran. Global stability of genetic systems governed by mutation and selection. *Mathematical Proceedings Cambridge Philosophical Society*, 80:331–336, 1976.

[128] P. A. P. Moran. Global stability of genetic systems governed by mutation and selection II. *Mathematical Proceedings Cambridge Philosophical Society*, 81:435–441, 1977.

[129] J.H. Moss, D.A. Beauchamp, A.D. Cross, K.W. Myers, E.V. Farley, J.M. Murphy, and J.H. Helle. Evidence for size-selective mortality after the first summer of ocean growth by pink salmon. *Transactions of the American Fisheries Society*, 134(5):1313–1322, 2005.

[130] Hans-Georg Müller, Jane-Ling Wang, James R. Carey, Edward P. Caswell-Chen, Carl Chen, Nikos Papadopoulos, and Fang Yao. Demographic window to aging in the wild: Constructing life tables and estimating survival functions from marked individuals of unknown age. *Aging Cell*, 3:125131, 2004.

[131] Stephan B. Munch and Marc Mangel. Evaluation of mortality trajectories in evolutionary biodemography. *Proceedings of the National Academy of Sciences, USA*, 103(44):1660416607, 2006.

[132] S. D. Mylius and O. Diekmann. On evolutionarily stable life-histories, optimization and the need to be specific about density-dependence. *Oikos*, 74:218–224, 1995.

[133] M. W. Nachman and S. L. Crowell. Estimate of the mutation rate per nucleotide in humans. *Genetics*, 156:297–304, 2000.

[134] M. Nakaoka. Size-dependent reproductive traits of *Yoldia notabilis* (bivalvia, protobranchia). *Marine Ecology Progress Series*, 114(1-2):129–137, 1994.

[135] M. Nakaoka. Size-dependent survivorship of the bivalve *Yoldia notabilis* (Yokoyama, 1920): The effect of crab predation. *Journal of Shellfish Research*, 15(2):355–362, 1996.

[136] A. Nault and D. Gagnon. Ramet demography of *Allium tricoccum*, a spring ephemeral, perennial forest herb. *Journal of Ecology*, 81(1):101–119, 1993.

[137] Sean Nee, Nick Colegrave, Stuart A. West, and Alan Grafen. The illusion of invariant quantities in life histories. *Science*, 309:1236–1239, 2005.

[138] Randolph M. Nesse. Life table tests of evolutionary theories of senescence. *Experimental Gerontology*, 23(6):445453, 1988.

[139] T. Noda. Shell growth of the sand snail *Umbonium costatum* (Kiener) in hakodate bay. *Bulletin of the faculty of Fisheries, Hokkaido University*, 42(4):115–125, 1991.

[140] T. Noda, S. Nakao, and S. Goshima. Life history of the temperate subtidal gastropod *Umbonium Costatum*. *Marine Biology*, 122(1):73–78, 1995.

[141] H. T.J. Norton. Natural selection and Mendelian variation. *Proceedings of the London Mathematic Society*, 28:1–45, 1928.

[142] T. Ohta and M. Kimura. A model of mutation appropriate to estimate the number of electrophoretically detectable alleles in a finite population. *Genetic Research*, 22:201–204, 1973.

[143] S. H. Orzack and S. Tuljapurkar. Population dynamics in variable environments. VII. The demography and evolution of iteroparity. *American Naturalist*, 133:901–923, 1989.

[144] M. A. Ottinger, R. E. Ricklefs, and C. E. Finch, editors. 2nd *Symposium on Organisms with Slow Aging (SOSA-2)*, volume 38(7) of *Experimental Gerontology (Special Issue)*, 2003.

[145] Tony L. Parkes, Andrew J. Elia, Dale Dickinson, Arthur J. Hilliker, John P. Phillips, and Gabrielle L. Boulianne. Extension of *Drosophila* lifespan by overexpression of human SOD1 in motorneurons. *Nature Genetics*, 19:171 – 174, 1998.

[146] Linda Partridge. Evolutionary biology and age-related mortality. In Kenneth W. Wachter and Caleb E. Finch, editors, *Between Zeus and the Salmon*, pages 78–95, Washington, D.C., 1997. National Academy Press.

[147] Linda Partridge. Evolutionary theories of aging applied to long-lived organisms. *Experimental Gerontology*, 36:641–650, 2001.

[148] Linda Partridge and Nicholas H. Barton. Optimality, mutation and the evolution of ageing. *Nature*, 362:305–311, 1993.

[149] Linda Partridge and Nicholas H. Barton. On measuring the rate of ageing. *Proceedings of the Royal Society London B*, 263:1365–1371, 1996.

[150] Linda Partridge and David Gems. The evolution of longevity. *Current Biology*, 12(16):R544–R546, 2002.

[151] Linda Partridge and Marc Mangel. Messages from mortality: the evolution of death rates in the old. *TREE*, 14:438–442, 1999.

[152] N. Perrin. Optimal resource allocation and the marginal value of organs. *American Naturalist*, 139(6):1344–1369, 1992.

[153] Nicolas Perrin and Richard M. Sibly. Dynamic models of energy allocation and investment. *Annual Review of Ecology and Systematics*, 24:379–410, 1993.

[154] I. Peterson and J.S. Wroblewski. Mortality rate of fishes in the pelagic ecosystem. *Canadian Journal of Fisheries and Aquatic Sciences*, 41(7):1117–1120, 1984.

[155] S. D. Pletcher, D. Houle, and J. W. Curtsinger. Age-specific properties of spontaneous mutations affecting mortality in *Drosophila Melanogaster*. *Genetics*, 148:287–303, 1998.

[156] Scott D. Pletcher, Stuart J. Macdonald, Richard Marguerie, Ulrich Certa, Stephen C. Stearns, David B. Goldstein, and Linda Partridge. Genome-wide transcript profiles in aging and calorically restricted *Drosophila melanogaster*. *Current Biology*, 12:712–723, 2002.

[157] L.S. Pontryagin. *The Mathematical Theory of Optimal Processes*. Wiley Interscience, New York, 1962.

[158] D. E. L. Promislow and M. Tatar. Mutation and senescence: Where genetics and demography meet. *Genetica*, 102/103:299–314, 1998.

[159] Daniel Promislow. Senescence in natural populations of mammals: A comparative study. *Evolution*, 45(8):1869–1887, 1991.

[160] Daniel E. L. Promislow. Protein networks, pleiotrophy and the evolution of senescence. *Proceedings of the Royal Society London, B*, 271:1225–1234, 2004.

[161] Daniel E.L. Promislow, Marc Tatar, A. A. Khazaeli, and J. W. Curtsinger. Age-specific patterns of genetic variance in *Drosophila melanogaster*. I. Mortality. *Genetics*, 143:839–848, 1996.

[162] Mark Rees, Andy Sheppard, David Briese, and Marc Mangel. Evolution of size-dependent flowering in *Onopordum illyricum*: A quantitative assessment of the role of stochastic selection pressures. *The American Naturalist*, 154, 1999.

[163] David N. Resznick, Michael J. Bryant, Derek Roff, Cameron K. Ghalambor, and Dionna E. Ghalambor. Effect of extrinsic mor-

tality on the evolution of senescence in guppies. *Nature*, 431:1095–1099, 2004.

[164] Robert E. Ricklefs. Evolutionary theories of aging: Confirmation of a fundamental prediction, with implications for the genetic basis and evolution of life span. *American Naturalist*, 152:24–44, 1998.

[165] Robert E. Ricklefs and Alex Scheuerlein. Biological implications of the Weibull and Gompertz models of aging. *Journal of Gerontology*, 57A(2):B69–B76, 2002.

[166] Bonnie J. Ripley and Hal Caswell. Recruitment variability and stochastic population growth of the soft-shell clam, *Mya arenaria*. *Ecological Modelling*, 2005.

[167] Deborah A. Roach and Jutta Gampe. Age-specific demography in *Plantago*: Uncovering age-dependent mortality in a natural population. *American Naturalist*, 164(1):60–69, 2004.

[168] Arthur J. Robson and Hillard S. Kaplan. Why we grow large and then grow old: Economics, biology and mortality. unpublished, 2005.

[169] Derek A. Roff. *Life History Evolution.* Sinauer Associates, Sunderland, Massachusetts, 2002.

[170] B. Rogina, R. A. Reenan, S.P. Nilsen, and S.L. Helfand. Extended life-span conferred by cotransporter gene mutations in drosophila. *Science*, 290:2137–2140, 2000.

[171] Karen Rose, Mark Rees, and Peter Grubb. Evolution in the real world: Stochastic variation in the determinants of fitness in *Carlina Vulgaris*. *Evolution*, 56:14161430, 2002.

[172] R. Michael Rose. *Evolutionary Biology of Aging.* Oxford University Press, New York, 1991.

[173] J. Sarukhn, M. Martinez-Ramos, and D. Pinero. *Perspectives on plant population ecology*, chapter The analysis of demographic variability at the individual level and its population consequences, pages 83–106. Sinauer Associates, Sunderland, Massachusetts, USA, 1984.

[174] J.R. Sauer and N.A. Slade. Uinta ground squirrel demography: Is body mass a better categorical variable than age? *Ecology*, 68:642–650, 1987.

[175] William M. Schaffer. Selection for optimal life histories: The effects of age structure. *Ecology*, 55(2):291–303, 1974.

[176] William M. Schaffer. The application of optimal control theory to the general life history problem. *American Naturalist*, 121(3):418–431, 1983.

[177] Stewart T. Schultz and Michael Lynch. Mutation and extinction: The role of variable mutational effects, synergistic epistasis, beneficial mutations, and degree of outcrossing. *Evolution*, 51(5):1363–1371, 1997.

[178] Carla M. Sgr and Linda Partridge. Evolutionary responses of the life history of wild-caught drosophila melanogaster to two standard methods of laboratory culture. *The American Naturalist*, 156:341353, 2000.

[179] F.R. Sharpe and Alfred J. Lotka. A problem in age-distribution. *Philosophical Magazine, Series 6*, 21:435–438, 1911.

[180] David R. Shook, A. Brooks, and Thomas E. Johnson. Mapping quantitative trait loci affecting life history traits in the nematode *Caenorhabditis elegans*. *Genetics*, 142:801–817, 1996.

[181] Marian Smith, Hal Caswell, and Paige Mettler-Cherry. Stochastic flood and precipitation regimes and the population dynamics of a threatened floodplain plant. *Ecological Applications*, 15(3):1036–1052, 2005.

[182] O.T. Solbirg. Studies on the population biology of the genus Viola. II. The effect of plant size on fitness in *Viola sororia*. *Evolution*, 35:1080–1093, 1981.

[183] Peter D. Sozou and Robert M. Seymour. To age or not to age. *Proceedings of the Royal Society London, B*, 271:457463, 2004.

[184] C.C. Spencer and D. E. L. Promislow. Age-specific changes in epistatic effects on mortality rate in *Drosophila melanogaster*. *Journal of Heredity*, 96(5):513–521, 2005.

[185] Christine C. Spencer, Christine E. Howell, Amber R. Wright, and Daniel E. L. Promislow. Testing an 'aging gene' in long-lived *Drosophila* strains: Increased longevity depends on sex and genetic background. *Aging Cell*, 2:123–130, 2003.

[186] Stephen C. Stearns. *The Evolution of Life Histories*. Oxford University Press, Oxford, 1992.

[187] D. Steinsaltz, S. N. Evans, and K. W. Wachter. A generalized model of mutationselection balance with applications to aging. *Advances in Applied Mathematics*, 35:16–33, 2005.

[188] Bernhard L. Strehler. *Time, cells and aging*. New York [et al.] : Academic Press, 1977.

[189] Marc Tatar, James R. Carey, and James W. Vaupel. Long term cost of reproduction with and without accelerated senescence in *Callosobruchus maculatus*: Analysis of age-specific mortality. *Evolution*, 47:1302–1312, 1993.

[190] Marc Tatar, A. Kopelman, D. Epstein, M.-P. Tu, C.-M Yin, and R. S. Garofalo. A mutant *Drosophila* insulin receptor homolog that extends life-span and impairs neuroendocrine function. *Science*, 292:107 – 110, 2001.

[191] Marc Tatar, Daniel E.L. Promislow, A. A. Khazaeli, and J. W. Curtsinger. Age-specific patterns of genetic variance in *Drosophila melanogaster*. II. Fecundity and its genetic covariance with age-specific mortality. *Genetics*, 143:849–858, 1996.

[192] Howard M. Taylor, Robert S. Gourley, Charles E. Lawrence, and Robert S. Kaplan. Natural selection of life history attributes: An analytical approach. *Theoretical Population Biology*, 5:104–122, 1974.

[193] A.T. Teriokhin. Evolutionarily optimal age schedule of repair: Computer modelling of energy partition between current and future survival and reproduction. *Evolutionary Ecology*, 12:291–307, 1998.

[194] D.W. Tinkle and R.E. Ballinger. *Sceloporus Undulatus*: Study of the intraspecific comparative demography of a lizard. *Ecology*, 53(4):570–584, 1972.

[195] D.W. Tinkle, A.E. Dunham, and J.D. Congdon. Life history and demographic variation in the lizard *Sceloporus graciosus*: A long-term study. *Ecology*, 74(8):2413–2429, 1993.

[196] Justin M. J. Travis. The evolution of programmed death in a spatially structured population. *Journal of Gerontology*, 59A(4):301–305, 2004.

[197] Meng-Ping Tu, Diane Epstein, and Marc Tatar. The demography of slow aging in male and female *Drosophila* mutant for the insulin-receptor substrate homolog chico. *Aging Cell*, 1:75–80, 2002.

[198] Shripad Tuljapurkar. The evolution of senescence. In Kenneth W. Wachter and Caleb E. Finch, editors, *Between Zeus and the Salmon*, pages 65–77, Washington, D.C., 1997. National Academy Press.

[199] Shripad D. Tuljapurkar. *Population Dynamics in Variable Environments, Lecture Notes in Biomathematics.* Springer Verlag, New York, 1990.

[200] J. W. Vaupel, A. Baudisch, M. Dölling, D. A. Roach, and J. Gampe. The case for negative senescence. *Journal of Theoretical Population Biology*, 65:339–351, 2004.

[201] J. W. Vaupel, J. R. Carey, K. Christensen, T. E. Johnson, A. I. Yashin, N. V. Holm, I. A. Iachine, V. Kannisto, A. A. Khaza-

eli, P. Liedo, V. D. Longo, Y. Zeng, K. G. Manton, and J. W. Curtsinger. Biodemographic trajectories of longevity. *Science*, 280:855–860, 1998.

[202] James W. Vaupel. Trajectories of mortality at advanced ages. In Kenneth W. Wachter and Caleb E. Finch, editors, *Between Zeus and the Salmon*, pages 17–37, Washington, D.C., 1997. National Academy Press.

[203] James W. Vaupel. Post-Darwinian longevity. In James R. Carey and Shripad Tuljapurkar, editors, *Life Span. Evolutionary, Ecological, and Demographic Perspectives*, pages 258–269, New York, 2003. Population Council.

[204] James W. Vaupel and James R. Carey. Compositional interpretations of medfly mortality. *Science*, 260:1666–1667, 1993.

[205] James W. Vaupel, James R. Carey, and Kaare Christensen. Its never too late. *Science*, 301:1679–1680, 2003.

[206] James W. Vaupel, Kenneth G. Manton, and Eric Stallard. The impact of heterogeneity in individual frailty on the dynamics of mortality. *Demography*, 16:439–454, 1979.

[207] James W. Vaupel and Anatoli I. Yashin. Heterogeneity's ruses: Some surprising effects of selection on population dynamics. *The American Statistician*, 39(3):176–185, 1985.

[208] Kenneth W. Wachter. Evolutionary demographic models for mortality plateaus. *Proceedings of the National Academy of Sciences, USA*, 96:10544–10547, 1999.

[209] Joshua Weitz and Hunter B. Fraser. Explaining mortality rate plateaus. *Proceedings of the National Academy of Sciences, USA*, 98(26):15383–15386, 2001.

[210] Geoffrey B. West, James H. Brown, and B. Enquist. A general model for ontogenetic growth. *Nature*, 413:628–631, 2001.

[211] Michael C. Whitlock and Denis Bourguet. Factors affecting the genetic load in *Drosophila*: Synergistic epistasis and correlations among fitness components. *Evolution*, 54(5):1654–1660, 2000.

[212] George C. Williams. Pleiotropy, natural selection, and the evolution of senescence. *Evolution*, 11(4):398–411, 1957.

[213] George C. Williams. Natural selection, the costs of reproduction, and a refinement of Lack's principle. *American Naturalist*, 100(916):687–690, 1966.

[214] Paul D. Williams and Troy Day. Antagonistic pleiotropy, mortality source interactions, and the evolutionary theory of senescence. *Evolution*, 57(7):1478–1488, 2003.

[215] D.L. Wilson. The analysis of survival (mortality) data: Fitting Gompertz, Weibull, and logistic functions. *Mechanisms of Ageing and Development*, 74:15–33, 1994.

[216] Lev Yu. Yampolsky, Louise E. Pearse, and Daniel E.L. Promislow. Age-specific effects of novel mutations in *Drosophila melanogaster*. I. Mortality. *Genetica*, 110:1129, 2001.

[217] Anatoli I. Yashin, Giovanna De Benedictis, James W. Vaupel, Kirill F. Andreev, Qihua Tan, Ivan A. Iachine, Massimiliano Bonaf, Maria De Luca, S. Valensin, L. Carotenuto, and Claudio Franceschi. Genes, demography, and life span: the contribution of demographic data in genetic studies on aging and longevity. *American Journal of Human Genetics*, 65(4):1178–1193, 1999.

[218] Anatoli I. Yashin, G. De Benedicts, James W. Vaupel, Qihua Tan, K. F. Andreev, Ivan A. Iachine, M. Bonafe, S. Valensin, M. De Luca, L. Carotenuto, and C. Franceschi. Genes and longevity: lessons from studies of centenarians. *Journal of Gerontology: Biological Sciences*, 55:B319–B328, 2000.

[219] Yodzis. Concerning the sense in which maximizing fitness is equivalent to maximizing reproductive value. *Ecology*, 62(6):1681–1682, 1981.

[220] Kengo Yoshidaa, Toshitaka Fujisawab, Jung Shan Hwanga, Kazuho Ikeoa, and Takashi Gojoboria. Degeneration after sexual differentiation in hydra and its relevance to the evolution of aging. *Gene*, 385:64–70, 2006.

Index